Advances in Intelligent Systems and Computing

Volume 684

Series editor

Janusz Kacprzyk, Polish Academy of Sciences, Warsaw, Poland
e-mail: kacprzyk@ibspan.waw.pl

The series "Advances in Intelligent Systems and Computing" contains publications on theory, applications, and design methods of Intelligent Systems and Intelligent Computing. Virtually all disciplines such as engineering, natural sciences, computer and information science, ICT, economics, business, e-commerce, environment, healthcare, life science are covered. The list of topics spans all the areas of modern intelligent systems and computing.

The publications within "Advances in Intelligent Systems and Computing" are primarily textbooks and proceedings of important conferences, symposia and congresses. They cover significant recent developments in the field, both of a foundational and applicable character. An important characteristic feature of the series is the short publication time and world-wide distribution. This permits a rapid and broad dissemination of research results.

More information about this series at http://www.springer.com/series/11156

Thanaruk Theeramunkong
Rachada Kongkachandra
Thepchai Supnithi
Editors

Advances in Natural Language Processing, Intelligent Informatics and Smart Technology

Selected Revised Papers from the Eleventh International Symposium on Natural Language Processing (SNLP-2016) and the First Workshop in Intelligent Informatics and Smart Technology, 10-12 February 2016, Phranakhon, Si Ayutthaya, Thailand

 Springer

Editors
Thanaruk Theeramunkong
School of Information, Computer, and
  Communication Technology (ICT)
Sirindhorn International Institute of
  Technology (SIIT)
Thammasat University
Khlong Luang, Pathum Thani
Thailand

Thepchai Supnithi
Language and Semantic Technology
  Laboratory
National Electronics and Computer
  Technology Center
Khlong Luang, Pathum Thani
Thailand

Rachada Kongkachandra
Faculty of Science and Technology
Department of Computer Science
Thammasat University
Khlong Luang, Pathum Thani
Thailand

ISSN 2194-5357   ISSN 2194-5365   (electronic)
Advances in Intelligent Systems and Computing
ISBN 978-3-030-09925-1   ISBN 978-3-319-70016-8   (eBook)
https://doi.org/10.1007/978-3-319-70016-8

Preface

The Eleventh International Symposium on Natural Language Processing (SNLP-2016) is the tenth workshop of the series of SNLP, biannually since 1993 with the cooperative effort of several universities in Thailand. The purpose of SNLP is to promote research in natural language processing and related fields. As a unique, premier meeting of researchers, professionals, and practitioners, SNLP provides a place and opportunity for them to discuss various current and advanced issues of interests in NLP. The eleventh conference will be held in 2016 and is hosted by Thammasat University in cooperation with the National Electronics and Computer Technology Center (NECTEC) and Artificial Intelligence Association of Thailand (AIAT). This conference offers excellent networking opportunities to participants, with a wonderful taste of local culture. High-quality research papers and case studies are invited to be submitted electronically through the conference's Web site.

This year, we expand the symposium to organize the First Workshop in Intelligent Informatics and Smart Technology. There are 94 papers submitted to all workshops. Sixty-six papers are accepted to present in the workshops.

After an intense discussion during the workshops, we selected papers from all accepted papers with 33% acceptance ratio. Therefore, this volume includes twelve regular papers from the Symposium on Natural Language Processing track and ten papers from Workshop in Intelligent Informatics and Smart Technology track. We believe that the workshops provide a valuable venue for researchers to share their works and could collaborate with like-minded individuals. We hope that readers will find the ideas and lessons presented in the proceedings relevant to their research.

Finally, we would like to thank the Eleventh International Symposium on Natural Language Processing (SNLP-2016) Executive Committees and Program

Co-Chairs for entrusting us with the important task of chairing the workshop program, thus giving us an opportunity to grow through valuable academic learning experiences. We also would like to thank all workshop co-chairs for their tremendous and excellent work.

Khlong Luang, Thailand   Thanaruk Theeramunkong
February 2016   Rachada Kongkachandra
                                                        Thepchai Supnithi

Organization

Organizing Committee

Honorary Co-Chairs

Pakorn Sermsuk   Thammasat University, Thailand
Sarun Sumriddetchkajorn   NECTEC, Thailand

General Co-Chairs

Asanee Kawtrakul   Kasetsart University, Thailand
Thanaruk Theeramunkong   Thammasat University, Thailand
Vilas Wuwongse   Asian University, Thailand
Virach Sornlertlamvanich   Thammasat University, Thailand

Steering Committee

Ir. Hammam Riza   Agency for the Assessment and Application
    of Technology (BPPT), Indonesia
Enya Kong Tang   Linton University College, Malaysia
Rusli bin Abd Ghani   Menara DBP, Malaysia
Rachel Edita O. Roxas   National University, Philippines
Ai Ti Aw   Institute for Infocomm Research, Singapore
Luong Chi Mai   Institute of Information Technology
    (IOIT), Vietnam

Joel Paz Ilao   De La Salle University, Philippines
Noy Shoung   National Institute of Posts, Telecoms and
                                          ICT (NIPTICT), Ministry of Post and
                                          Telecoms, Cambodia

Program Committee

Alisa Kongthon   NECTEC, Thailand
Chalermsub Sangkavichitr   King Mongkut's University of
                                          Technology Thonburi, Thailand
Chutamanee Onsuwan   Thammasat University, Thailand
Kanyalag Phodong   Thammasat University, Thailand
Kazuhiro Takeuchi   Osaka EC University, Japan
Kreangsak Tamee   Naresuan University, Thailand
Mahasak Ketcham   King Mongkut's University of Technology
                                          North Bangkok, Thailand
Makoto Okada   Osaka Prefecture University, Japan
Marut Buranarach   NECTEC, Thailand,
Masaki Murata   Tottori University, Japan
Narit Hnoohom   Mahidol University, Thailand
Narumol Chumuang   Muban Chombueng Rajabhat University, Thailand
Pakorn Leesutthipornchai   Thammasat University, Thailand
Pokpong Songmuang   Thammasat University, Thailand
Pongsagon Vichitvejpaisal   Thammasat University, Thailand
Ponrudee Netisopakul   King Mongkut's University of Technology
                                          Ladhrabang, Thailand
Prachya Boonkwan   NECTEC, Thailand
Sanparith Marukatat   NECTEC, Thailand
Sunee Pongpinigpinyo   Silpakorn University, Thailand
Supaporn Kiattisin   Mahidol University, Thailand
Thepchai Supnithi   NECTEC, Thailand
Virach Sornlertlamvanich   Thammasat University, Thailand
Wasit Limprasert   Thammasat University, Thailand

Contents

Part I

SNLP 2016: The Symposium on Natural Language Processing

Semi-automatic Framework for Generating RDF Dataset from Open Data

Pattama Krataithong[1,2(✉)], Marut Buranarach[1(✉)],
Nuttanont Hongwarittorrn[2], and Thepchai Supnithi[1]

[1] Language and Semantic Technology Laboratory, National Electronics and
Computer Technology Center (NECTEC), Pathumthani, Thailand
{pattama.kra,marut.bur,thepchai.sup}@nectec.or.th
[2] Department of Computer Science, Faculty of Science and Technology,
Thammasat University, Pathumthani, Thailand
nth@cs.tu.ac.th

Abstract. Most of datasets in open data portals are mainly in tabular format in spreadsheet, e.g. CSV and XLS. To increase the value and reusability of these datasets, the datasets should be made available in RDF format that can support better data querying and data integration. In this paper, we present a semi-automatic framework for generating and publishing RDF dataset from existing datasets in tabular format. This framework provides automatic schema detection functions, i.e. data type and nominal type detection. In addition, user can create RDF dataset from existing dataset step-by-step without required knowledge about RDF and OWL. Evaluation of the schema detection using some datasets from data.go.th shows that the technique can achieve high precision in most datasets and high recall for the datasets with small data input errors.

Keywords: Open data publishing process · Automatic schema detection · RDF dataset publishing

1 Introduction

Open Government Data initiatives are widely adopted in many countries in order to increase the transparency, public engagement and the channel of communication between government and their citizens. In Thailand, the government has promoted the initiative by issuing policies and regulations related to open data publishing and educating government sectors to understand about open government data. Data.go.th is the national open government data portal of Thailand, which started to publish several datasets in 2014. As of December 2015, there are over four hundred datasets available on this portal in various formats (e.g. XLS, CSV, PDF) and users can download raw datasets from this web site. Data.go.th also promotes applications and innovations that are created based on the content published on the portal.

© Springer International Publishing AG 2018
T. Theeramunkong et al. (eds.), *Advances in Natural Language Processing,
Intelligent Informatics and Smart Technology*, Advances in Intelligent Systems

Based on the 5-star open data model,[1] Resource Description Framework (RDF) is a standard data format that can support linked open data. There are two important standards for integrating data. First, RDF is a standard format for integrating data based on URI and XML syntax.[2] Second, the Web Ontology Language (OWL)[3] is important for linked data based on classes and properties. Datasets on open data portal are continually increasing. Majority of datasets on Data.go.th are in tubular formats such as Excel and CSV that are defined as the level 2 and 3 of the open data model. To support integrating data, the datasets need to be converted to RDF format. Therefore, we present a semi-automatic framework for generating and publishing RDF dataset on the open data portal. Figure 1 shows the RDF dataset publishing workflow.

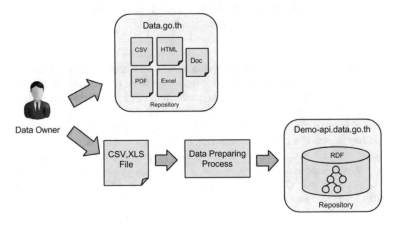

Fig. 1. Dataset publishing workflow

In this paper, we describe a semi-automatic framework in transforming tabular data in spreadsheet format to RDF format. This framework provides the user support system for creating RDF dataset and OWL file step-by-step that allows data owner to publish a dataset by themselves. Data owner can publish their dataset easily and rapidly, although they do not have the knowledge about RDF and OWL. This paper also focuses on techniques for data type and nominal type detection for column fields in each dataset. Data type and nominal type detection is an important step to automatic ontology creation and generating linked RDF datasets. Specifically, columns with nominal type are transformed to "ObjectProperty" in OWL and non-nominal type are transformed to "DatatypeProperty" with respective data types in OWL. Our technique of the nominal type detection for text and numeric data columns are presented. Evaluation results of the technique conducted with 13 datasets are provided.

---

[1] http://5stardata.info/en/.

[2] http://www.w3.org/RDF/.

[3] http://www.w3.org/2001/sw/wiki/OWL.

2   A Motivating Example

Schema of datasets in open data portal is important for integrating and analyzing data. In this section, we present an example that motivates the need of automatic schema detection before converting a tabular dataset to RDF and OWL. The comparison between non-detection and detection of the schema are as follows:

No Schema Detection: the most common form of generating RDF is without schema detection. The direct mapping is the automatic approach to convert a dataset with no need for inputs from users. This approach is the standard for converting relational database to RDF, which is provided by the W3C RDB2RDF Working Group. This approach takes three inputs including database, set of primary keys and foreign key [1]. Adopting this approach to converting tabular data to RDF, a table is usually mapped to ontology class and each column is mapped to a datatype property that has literal values, so each dataset is independent and not linked with others.

Schema Detection: there are several possible data types in each dataset such as string, integer, and float. Data type detection is the first step for finding the semantic type. Each data type can be categorized to various types including nominal, ordinal, interval and ratio types. We are interested in the nominal types because most of these types can refer to semantic types. For example, data types of reservoir statistics dataset consist of string, integer, and float. These string types can be categorized to date and nominal types, as shown in Fig. 2.

	Date	Nominal	Integer					Float	
statID	date	damName	capacity	currentWater	currentWaterPercentage	usableWater	usablePercentage	ingressWater	egressWater
1	01-01-11	เขื่อนภูมิพล	13462	8281	62	4481	33	3.21	14
2	01-01-11	เขื่อนสิริกิติ์	9510	7440	78	4590	48	7.33	25
3	01-01-11	เขื่อนแม่งัด	265	269	102	247	93	0	0.39
4	01-01-11	เขื่อนกิ่วลม	112	100	89	96	86	0.29	1.25
5	01-01-11	เขื่อนแม่กวง	263	160	61	146	56	0.22	0.03
6	01-01-11	เขื่อนกิ่วคอหมา	170	176	104	170	100	0.15	0.15
7	01-01-11	เขื่อนแควน้อย	769	682	89	646	84	1.75	3.46
8	01-01-11	เขื่อนลำปาว	1430	1166	82	1081	76	1.1	5.3
9	01-01-11	เขื่อนลำตะคอง	314	336	107	309	98	0.46	1.22
10	01-01-11	เขื่อนลำพระเพลิง	110	106	96	105	95	0.02	0.02
11	01-01-11	เขื่อนน้ำอูน	520	308	59	265	51	0	0.08
12	01-01-11	เขื่อนอุบลรัตน์	2432	2067	85	1486	61	3.55	5.43
13	01-01-11	เขื่อนสิรินธร	1966	1527	78	696	35	0.78	0
14	01-01-11	เขื่อนจุฬาภรณ์	164	150	91	106	65	0	0.48
15	01-01-11	เขื่อนห้วยหลวง	118	109	92	104	88	0	0.57
16	01-01-11	เขื่อนน้ำพรอง	121	85	70	82	68	0.1	0.02
17	01-01-11	เขื่อนมูลบน	141	116	82	109	77	0	0
18	01-01-11	เขื่อนน้ำพุง	165	115	70	106	64	0	0.56
19	01-01-11	เขื่อนลำแชะ	275	244	89	237	86	0	0
20	01-01-11	เขื่อนป่าสักฯ	960	771	80	768	80	0.78	3.03
21	01-01-11	เขื่อนกระเสียว	240	248	103	208	87	0.24	0.06
22	01-01-11	เขื่อนทับเสลา	160	102	64	94	59	0	1.65
23	01-01-11	เขื่อนศรีนครินทร์	17745	14076	79	3811	21	3.43	6.14

Fig. 2. Schema detection

There are two data types that are necessary for detection. First, we attempt to categorize "string" type to literal values or nominal types. Nominal types are usually in string data type [2]. These types can be grouped and transformed to object property in ontology class and refered to resource with an URI. Sometimes these types can be denoted with number that does not mean quantity. For example, we refer to province name with province code, e.g. "Bangkok" = 1, "Chiang Mai" = 2, etc. Otherwise, "numeric" type is typically important for statistics analysis or for math operations e.g. add, subtract, multiple, and divide. To detect numeric values, we categorize these types to integer, float, and nominal types.

Some advantages of automatic schema detection are as follows:

(1) Detection of nominal types is the basis important step to integrate many datasets because this step will lead to alignment of various resources to use the same URI for identification of common things.
(2) The correctness of data types detection both string and numeric can increase the efficiency of recommended data analysis.

3 Generating CSV to RDF Dataset Workflow

This framework takes one input: a tabular data such as CSV, Excel. Then it generates RDF and OWL files from the tabular data. The RDF dataset generations consist of four processes: (1) User Management and Authentication (2) Dataset Preparing and Import (3) Schema Detection and Verification and (4) OWL and RDF data generation.

The requirements of input data are as follows: (1) the dataset must consist of only one table (one spreadsheet), (2) the table must have one header row, (3) header of the table must be written in English or Thai language.

The detailed workflow for the CSV to RDF data generation process is shown in the activity diagram in Fig. 3. The workflow for creating, updating, and deleting an RDF dataset is described as follows. The two involved entities are User and System. The process begins with the user authentication activity. After successful authentication, the user can begin to manage the dataset under his or her account. For creating a dataset, user must define the dataset information, upload CSV file, verify the detected schema, and transform the dataset. For updating a dataset, the system allows the user to upload new CSV file, verify schema, and retransform the updated dataset. Then, the system will create a new version of the dataset. There are a few requirements in updating a dataset as follows: First, the user cannot modify the dataset name. Second, the system allows the user to update a new CSV file only when the new file has the same schema. The update of existing schema, i.e. adding or removing fields or changing data types, is not permitted. This is primarily because schema update would change the API of the dataset, which could affect the applications that already used the dataset API. Thus, in our framework, the dataset with updated schema should be treated as a new dataset.

Figure 4 shows a design of the user interface that lists all datasets which a user has created. The user can choose to create, update, and delete the datasets. The system also allows the user to access each created dataset in two forms: viewing RDF and querying the dataset.

The following section describe the implementation details of the CSV to RDF data generation process.

3.1 User Management and Authentication

User registration module allows user account to be created before the user can publish a dataset. Organization name of each user must be defined and used as part of the default namespace for the user.

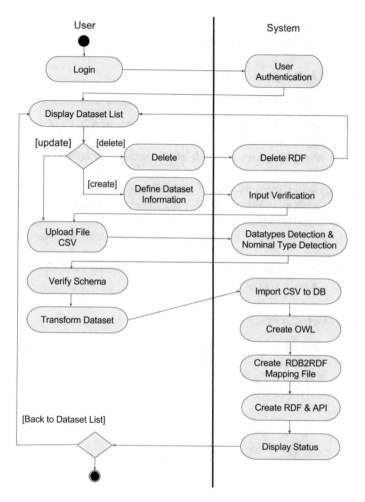

Fig. 3. Activity diagram of the CSV to RDF data generation process

Fig. 4. Design of the user interface that lists all datasets and possible actions

Normally, the user who can create RDF and OWL files is a data owner or representative officer who must be responsible for the metadata and content of the dataset.

3.2 Dataset Preparing and Import

The steps for preparing, identifying and uploading data are listed below:

- *Data Preparing*: User needs to prepare a tabular dataset in CSV or Excel format, which is published at the open data portal (Data.go.th). Next, the dataset must be cleaned and arranged to a structured form that should follow the canonical model of tabular data [3].
- *Data Identifying*: User must identify a dataset name and base namespace for a dataset. The system provide default namespace that follows the pattern "http:// demo-api.data.go.th/{organization}/{dataset_name}".
- The organization name of the user must be a unique value and must be written in English.
- *Data Uploading*: User needs to upload the dataset in CSV format to the Demo-api.data.go.th portal.

3.3 Schema Detection and Verification

In this step, the system will read a tabular file and auto-detect the schema. The data schema detection process consists of three steps including data type detection, nominal type detection, and semantic type detection, as shown in Fig. 5. Currently, only data type and nominal type detection are provided.

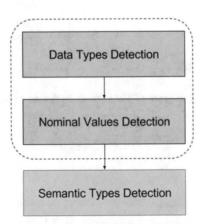

Fig. 5. Schema detection workflow

(1) *Data Type Detection*: the first step, the system will auto-detect the data types by using data in each column. Tabular data types can be categorized into two main groups that are string and numeric. String type can be categorized into three sub-groups including literal, date, and nominal types. Numeric type can be

categorized into three sub-groups including integer, float, and nominal types as shown in Fig. 6.

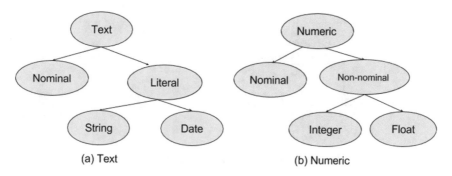

(a) Text   (b) Numeric

Fig. 6. Data type and nominal type detection

(2) *Nominal Type Detection*: nominal types are used to transform to object property in ontology class. These values are possible for both string and integer. Nominal type detection needs to use the information from data header, data types and cell values in each column.

We propose a nominal type detection method with simple formula to analyses each column by Unique value ratio (U), the formula is as follows:

$$Unique\ value\ ratio\ (U) = \frac{Number\ of\ unique\ values\ for\ the\ column}{Number\ of\ all\ records}$$

We set a threshold that for a column with nominal type, U must be less than 0.1 because a small ratio of number of unique values to number of all records indicating the column to have nominal types. The nominal type detection for text and numeric columns are slightly different. Numeric columns with some range values such as age, percentage, which normally contains only the values between 0 and 100, can be incorrectly detected as nominal type using only the U value. Thus, for numeric column, column header is also analyzed. For example, numeric data that has nominal type usually are only some ID numbers. The header texts of these columns normally includes terms such as ID, code, รหัส, เลขที่, หมายเลข. In addition, the column with the U value of 1 indicates that it contains unique IDs, i.e. primary key, which indicates nominal type. Nominal type detection flow as shown in Fig. 7.

(3) *Semantic Types Detection*: Semantic type detection is used to detect the class of values in the nominal columns, e.g. Person, Organization, etc. For example, the Karma system uses machine learning to learn the semantic type inputs from users. The system will learn and suggest semantic types for next users [4]. Semantic type detection is still under development in our framework.

Our framework requires the user's schema verification process in order to ensure the correctness of the generated data schema. The result will be displayed on user

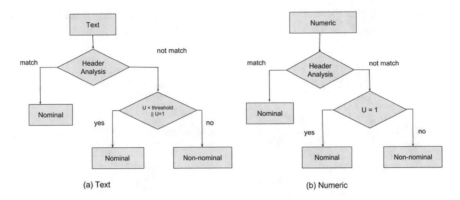

Fig. 7. Nominal type detection flow for text and numeric values

interface for user to verify the results as shown in Fig. 8. The output of this step is the schema of the dataset that will be used for the RDF and OWL files creation processes.

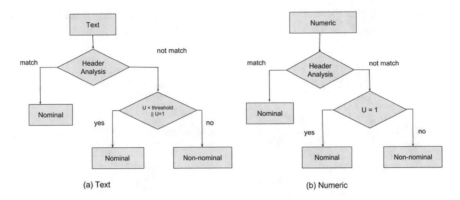

Fig. 8. Schema verifying interface

3.4   RDF and OWL Creation

In this step, we propose the processes for using the schema of the tabular data to generate OWL file and transform the tabular data to RDF dataset with a RDB to RDF mapping approach.

Database Creation: The purpose of this step is for importing tabular data to relational database. Database name and table name are the same as a dataset name. The system will create an auto increment field as a primary key for each dataset.

OWL File Creation: This step uses the results from the schema verification process to generating OWL file automatically.

Database Schema to Ontology Mapping: In this step, we explain an example of database-ontology mapping process. Consider the "expense statistics in 2002–2013" dataset from Data.go.th in CSV format, and then imported to relational database. Alter all labels of header table are renamed to "col1, col2, col3" and an auto increment ID is added as a primary key for a table as shown in Fig. 9. The schema of expense statistic database consists of only one table *"expense (Id, col1, col2, col3)"*, *"Id"* is a primary key, "col1" and "col2" are the nominal types, so cells values in each row are mapped to object properties and "col3" are mapped to datatype property in the ontology class.

RDF Dataset Creation: The purpose of this step is for transforming a tabular data to create RDF dataset using Apache Jena, D2RQ and TDB triples storage.

จังหวัด	ปี	รายจ่ายต่อครัวเรือน
กรุงเทพมหานคร	2545	21,919
สมุทรปราการ	2545	14,836
นนทบุรี	2545	22,702
ปทุมธานี	2545	17,836
พระนครศรีอยุธยา	2545	11,335
อ่างทอง	2545	8,834
ลพบุรี	2545	8,826

Object Properties ⬇ Datatype Property

Id	col1	col2	col3
1	กรุงเทพมหานคร	2545	21,919
2	สมุทรปราการ	2545	14,836
3	นนทบุรี	2545	22,702
4	ปทุมธานี	2545	17,836
5	พระนครศรีอยุธยา	2545	11,335
6	อ่างทอง	2545	8,834
7	ลพบุรี	2545	8,826

Fig. 9. Expense statistics dataset (database)

4 Evaluation

The schema detection algorithm was evaluated with 13 datasets from Data.go.th. The 13 tables consist of the total of 198 columns, there were 59 columns from 69 columns correctly detected as nominal type. The average precision for 59 columns was 0.98 and average recall was 0.84. The results of precision and recall of each dataset are shown in Table 1.

In nominal type detection, data input error is one of the most common problems. The errors include missing values, repeating row values, and filling data in wrong fields that can affect the effectiveness of the schema detection process. In data type detection, date detection is not effective when patterns of date are invalid. For example, date data

Table 1. The results of precision and recall of nominal type detection for each dataset

Dataset name	Detected nominal (columns)		Correct nominal (columns)	Total number of columns	Precision	Recall
	Correct detected	Total detected				
city_bus	5	5	7	7	1	0.71
tambon_latlong	10	10	10	12	1	1
accident_newyear	16	16	16	19	1	1
dam_stat	2	2	2	10	1	1
rainfall_stat	4	4	4	32	1	1
air_pullution	1	1	1	67	1	1
income_stat	2	2	2	3	1	1
expense_stat	2	2	2	3	1	1
factory_101_105_106	7	9	8	16	0.77	0.875
juristic_persons	2	2	3	7	1	0.66
gproject_may58	1	1	2	4	1	0.5
gbudget_may58	4	4	8	14	1	0.5
berth_thailand	3	3	4	4	1	0.75

does not has a separate character {*yyyymmdd*} or date data has only date {*dd*}, our algorithm will detect these data as integer type.

5 Related Work

There are several tools for generating RDF from tabular data. The related work involves transforming tabular to RDF and finding the semantic types approaches.

5.1 Transforming Tabular to RDF Approaches

The RDF123 is the open-source tool that can transform spreadsheet to RDF. This framework provides the flexible user interface for users to identifying the relationship of each column in spreadsheet. RDF graph is created based on directed labeled graph, vertexes are literals/resources and edges are the RDF triples [5]. This framework needs inputs from user for identifying the relationship in spreadsheet in order to create RDF graph, while our framework needs inputs from user for verifying schema and uses the verified schema for generating OWL and RDF dataset automatically.

The CSV2RDF is the semi-automatic conversion for CSV to RDF. This framework has a concept of mapping verification by data owners, community and consumers. Mapping approach is automatic default mapping based on Sparqlify-CSV, then transform tabular data to RDF and MediaWiki pages are created [3]. This framework can transform a tabular to RDF rapidly with default mapping without schema detection process. Our framework also focuses on schema detection process to increase the efficiency of recommended data analysis and data integration.

5.2   Finding Semantic Types Approach

The semantic type detection is still under development in our framework. There are many research approach related to finding semantic types as follows:

Varish Mulward at et. developed the automatic converting tabular data to RDF. They proposed the semantic types prediction algorithm that mapping between column headers and ontology class by using query to find the semantic types on Wikiology that using four vocabularies database consisting of DBpedia, Ontology, Freebase, WordNet, and Yago [6].

The Karma is the semi-automatic mapping between relational database and ontology to create RDF triples and update source model for semantic type suggestion. This framework uses machine learning to construct source model based on the data in each column and set of learning probabilistic of the semantic types from prior use [4].

Sande et al. [7] proposed the lightweight tabular data to RDF conversion. This approach using string, lexical, data types and context analysis for automatic mapping between column header and ontology. Each column are created to context graph and then uses Steiner tree algorithm to find optimal path in graph.

6   Conclusion and Future Work

We developed a semi-automatic framework for generating RDF dataset from open tabular data. This framework allows the users to publish their datasets in RDF format with no required knowledge about RDF and OWL. In this project, we focused on schema detection approach for data type and nominal type automatic detection. To improve the correctness of schema detection, users can verify and update the results of schema detection before the resulted schema is used in building RDF dataset. Our evaluation results with 13 datasets showed that the schema detection algorithm for nominal type can have a good precision and recall for cleaned tabular data in spreadsheet. Additional analysis and processing will be required to handle more complex cases and errors in the datasets.

In future work, we plan to use the nominal type detection results for finding the semantic types and to develop a semi-automatic RDF dataset linking tool.

Acknowledgements. This project was funded by the Electronic Government Agency (EGA) and the National Science and Technology Development Agency (NSTDA), Thailand.

References

1. Sequeda, J., Arenas, M., Miranker, D.P.: A completely automatic direct mapping of relational databases to RDF and OWL. In: The 10th International Semantic Web Conference (ISWC2011), October 2011
2. Types of data measurement scales: nominal, ordinal, intervalMy Market Research Methods. http://www.mymarketresearchmethods.com/types-of-data-nominal-ordinal-interval-ratio/. Accessed 23 Dec 2015

3. Ermilov, I., Auer, S., Stadler, C.: User-driven semantic mapping of tabular data. In: Proceedings 9th International Conference Semantanic Systems—I-SEMANTICS '13, vol. 105 (2013)
4. Knoblock, C.A., Szekely, P., Ambite, J.L., et al.: Semi-automatically mapping structured sources into the semantic web. In: Lecture Notes in Computer Science (including subseries Lecture Notes in Artificial Intelligence and Lecture Notes in Bioinformatics), pp. 375–390 (2012)
5. Han, L., Finin, T., Parr, C., et al.: RDF123: From spreadsheets to RDF. Lect. Notes Comput. Sci. (including Subser. Lect. Notes Artif. Intell. Lect. Notes Bioinformatics). **5318** LNCS, 451–466 (2008)
6. Mulwad, V., Finin, T., Joshi, A.: Automatically generating government linked data from tables. In: Working notes of AAAI Fall Symposium on Open Government Knowledge: AI Opportunities and Challenges, Nov 2011
7. Sande, M.V., De Vocht, L., Van Deursen, D., et al.: Lightweight transformation of tabular open data to RDF. In: 8th International Conference on Semantic Systems, pp. 38–42 (2012)

Design for a Listening Learner Corpus for a Listenability Measurement Method

Katsunori Kotani[1(✉)] and Takehiko Yoshimi[2]

[1] College of Foreign Studies, Kansai Gaidai Univeristy, Osaka, Japan
kkotani@kansaigaidai.ac.jp
[2] Faculty of Science and Technology, Ryukoku University, Shiga, Japan
yoshimi@rins.ryukoku.ac.jp

Abstract. A design for a listening learner corpus as a language resource for computer-assisted language learning systems is proposed, and a pilot learner corpus compiled from 20 university learners of English as a foreign language is reported. The learners dictated a news report. The corpus was annotated with part-of-speech tags and error tags for the dictation. The validity of the proposed learner corpus design was assessed on the basis of the pilot learner corpus, and was determined by examining the distribution of errors and whether the distribution properly demonstrated the learners' listening ability. The validity of the corpus was further assessed by developing a listenability measurement method that explains the ease of listening of a listening material for learners. The results suggested the dictation-based corpus data was useful for assessing the ease of listening English materials for learners, which could lead to the development of a computer-assisted language learning system tool.

Keywords: Learner corpus · Computer-assisted language learning · Listenability

1 Introduction

Computer-assisted language learning (CALL) systems have contributed to the successful learning and teaching of English as a foreign language (EFL) [1]. Current CALL systems are implemented by natural language processing techniques to evaluate learner language use. Development of effective CALL systems using statistical methods is derived from language resources, known as learner corpora, which demonstrate how learners use English; both correctly and incorrectly.

Although learner corpora are widely used for writing- or speaking-error analyses, use for listening-error analyses and listening material evaluation is warranted. Evaluation of listening materials means to assess the ease with which learners identify speech sounds, henceforth referred to as "listenability." If EFL learners are provided with materials in which the listenability matches the learner's individual proficiency, then their learning motivation increases, resulting in increased learning effects [2, 3].

To prepare listening materials with listenability appropriately matched for learner proficiency and interest, language teachers would need to evaluate the listenability of each material; however, such evaluation is time-consuming. Thus, it is plausible to

© Springer International Publishing AG 2018
T. Theeramunkong et al. (eds.), *Advances in Natural Language Processing,*
Intelligent Informatics and Smart Technology, Advances in Intelligent Systems

reduce teacher workload with a CALL system that measures the listenability of listening materials. A listening learner corpus for the development of a listenability measurement method needs to be annotated with tags for listening errors and linguistic properties, such as the part-of-speech.

From this requirement, a design was drafted for a listening learner corpus. Following the proposed corpus design, a preliminary listening learner corpus, henceforth referred to as a pilot learner corpus, was compiled. The validity assessment of the proposed corpus design, by examining whether the pilot learner corpus was useful for the development of a listenability measurement method, is reported. First, the corpus design was considered valid if the data demonstrated no extreme distribution, such as all of the words correctly listened to by the learners or no error tags included in the corpus, as data with such extreme distribution are useless for examining which linguistic properties affect listenability. Second, the corpus design was also considered valid if the corpus data demonstrated well-acknowledged listenability properties such as the influence of sentence length, where the longer a sentence is, the less listenability the sentence has; and presence of suffixes, as suffixes decreases the listenability since they are usually not clearly pronounced. Lastly, the corpus design was considered valid if a listenability measurement method based on the proposed corpus demonstrated acceptable performance.

2 Previous Studies

Previous learner corpus research focused on writing or speaking [4–6], but not too much on listening. To the best of the authors' knowledge, the following listening learner corpora have been compiled: a corpus by Kotani et al. [4] for EFL learners; a corpus by Rytting et al. [5] for learners of Arabic as a foreign language; and a corpus by ten Bosch et al. [6] for learners of Dutch as a second language.

2.1   Kotani et al. (2011)

The corpus developed by Kotani et al. [4], consisting of 127,170 words, was compiled to develop a listenability measurement method, and demonstrated listenability of EFL learners. Listenability was evaluated by 90 EFL learners. The learners listened to the materials and assigned listenability scores, based on their own subjective judgement, scored on a five-point Likert scale (1: very difficult; 2: difficult; 3: neutral; 4: easy; and 5: very easy). The listenability score was annotated on each sentence.

Learner proficiency was classified in terms of Test of English for International Communication (TOEIC) scores (mean score, 633.8; standard deviation, ±198.3) into three proficiency levels: beginner (n = 30), 280–485; intermediate (n = 30), 490–725, and advanced (n = 30) 730–985. The listening materials were four news reports taken from the Voice of America (VOA: http://www.voanews.com), a dynamic multimedia broadcaster funded by the U.S. Government that broadcasts accurate, balanced, and comprehensive news and information for international audiences.

Following the validation criteria of the present corpus design (see Sect. 1 for outline, and Sect. 3.4 for more details), the corpus design by Kotani et al. [4] was considered valid because the data reflected well-acknowledged listenability properties: sentence length influence, and learner proficiency influence. According to Kotani et al. [4], their corpus data demonstrated that sentence length affected listenability, and that proficient learners found the listenability to be high in a listening material that was judged as low listenability by non-proficient learners.

2.2 Rytting et al. (2014)

The corpus developed by Rytting et al. [5], consisting of 16,217 words, was compiled to develop a listening error-correction tool, and demonstrated the listenability of learners of Arabic as a foreign language. Listenability was evaluated based on four types of dictation error tags: deletion-type errors (such as deletion of a target phoneme); insertion-type errors (such as the addition of another phoneme to a target phoneme); substitution-type errors (such as substituting a target phoneme with another phoneme); and no-response type errors (where no phoneme is dictated for a target phoneme). The dictation error tags were annotated to each word.

In compiling this corpus, 62 Arabic learners wrote down an audio recording in Arabic that they heard. Learner proficiency was classified in terms of period of learning Arabic (mean period, 5.6 semesters; median, 4 semesters). The recording comprised 261 words, excluding basic words, and covered Arabic consonant sounds.

The corpus design by Rytting et al. [5] was considered valid because the data demonstrated no extreme distribution of the correct-dictation and no-response tags. According to Rytting et al. [5], 37.1 and 48.2% of all the words were correctly and incorrectly dictated, respectively. Although it was not mentioned, the remaining 14.7% was judged as no-response.

2.3 Ten Bosch et al. (2015)

The corpus developed by ten Bosch et al. [6], consisting of 1784 words, was compiled to investigate problems in listening comprehension and develop reference data for evaluating a computational model of human word recognition. This corpus was annotated with listenability scores for learners of Dutch as a second language. In this corpus, error tags were annotated on not all of the words, but on target words only. At least two target words appeared in each sentence of two listening materials, with each listening material comprising a story consisting of 11 sentences of 176 words. Listenability was classified with four types of error tags of dictation: deletion-type errors, substitution-type errors, insertion-type errors, and spelling-type errors.

In compiling this corpus, 58 learners wrote down an audio recording in Duetch that they heard. Learner proficiency was classified in terms of level of the Common European Framework of Reference for Languages, ranging from A2 level to B1 level.

The corpus design by ten Bosch et al. [6] was considered valid because the data demonstrated no extreme distribution on the correct-dictation and no-response tags. According to ten Bosch et al. [6], 67.4 and 32.6% of the target words were correctly and incorrectly dictated, respectively.

3   Design of Listening Learner Corpus

3.1   Participants

A learner corpus that is widely open to EFL learners and covers beginner-to-advanced levels exhaustively would be a useful resource for CALL. In order to measure listenability that suits learners at different levels, this study regarded a range of individual proficiency levels of EFL learners as a validity condition during the design of the present learner corpus. The proficiency of EFL learners will be determined with English test scores that measure the general English ability of EFL learners, such as TOEIC or Test of English as a Foreign Language (TOEFL). According to their test scores, EFL learners will be classified into at least three levels or proficiency.

Their educational background was also considered as a validity condition since appropriate listening materials differ depending on the educational backgrounds, such as those ranging from junior-high school to university. Among these educational backgrounds, this study determined to target EFL learners at university.

The present pilot learner corpus was compiled from 21 university EFL learners whose first language was Japanese. Their proficiency was relatively low ranging from the beginner- to intermediate-level in terms of TOEFL Institutional Program (IP) scores (range, 380–470; mean, 431.3; standard deviation, ±25.0).

The pilot corpus partially satisfied the proficiency condition requiring learners from beginner-to-advanced levels. In order to expand the present corpus, the corpus data from advanced-level learners must also be included.

3.2   Material

During the design of the present corpus study, listening material type was taken into consideration. Since English tests such as TOEIC and TOEFL often use (short) conversation as listening material, this study regarded conversation as plausible. However, news reports were chosen because they are concisely formed without hesitations or redundant pauses, unlike conversations; include current English vocabulary; and include proper noun phrases, which are less accessible in teaching materials designed for language learning.

The source of news reports was also taken into consideration for this study. Among various news sites, the VOA was chosen as the source material because the VOA provides two types of news reports; one for language learners with limited vocabulary and a decreased speech rate (approximately half to two-thirds of natural speech rate), and another for English speakers without any such restrictions. Thus, the VOA-based listening corpus was considered to enable further comparison of listenability between limited and non-limited vocabularies, and between slow and natural speech rates.

The VOA news report used in our pilot learner corpus consisted of 19 sentences as shown in Table 1. The mean number of words per sentence was 13.9 words, with a standard deviation of ±4.8. This news report was spoken at a speech rate of 67.6 words per minute, roughly half of natural speech rate. Proper nouns appeared in sentences (1), (2), (9), (15), (18) and (19).

Table 1. Sentences from the listening material

Number	Sentence
1	From VOA learning english, this is the education report
2	Lawmakers in Washington are debating educations issues, including the interest rates that students pay for loans
3	College students who take loans graduate owing an average of $26,000
4	But some economists say the real issue is controlling the cost of college
5	Experts say these high costs are hurting the whole economy
6	For the past 30 years, college tuition has been increasing at twice the rate of inflation
7	Universities say decreasing financial support from state governments forces them to charge higher tuition
8	On average, private colleges now charge more than $30,000 a year
9	Terry Hartle is a spokesman for the American Council on Education, which represents thousands of colleges across the United States
10	He says funding for education has been shrinking for years
11	Experts worry that the high cost of education makes it less likely that good students from poor families can attend college
12	This means fewer scientists, engineers and others who could help increase economic growth
13	Also, a survey shows that some students concerned about repaying thousands of dollars in loans are delaying marriage and children
14	Many cannot afford to invest in a house or buy a car
15	Georgetown University labor economist Anthony Carnevale says the current system cuts economic growth for the whole country
16	And the effects are important
17	He says meeting the demand for workers with higher education could add $500 billion to the American economy
18	A new government report says one-third of Americans age 25–29 now hold college degrees -up from one-fourth in 1995
19	For VOA learning english, I'm Mario Ritter

3.3   Procedure

This study aimed to collect corpus data based on the results of the listening task, from which the pilot corpus data was compiled. Following previous research [5, 6], a dictation task was chosen as it enables listenability analysis at the word-level. Word-level listenability analysis was expected to identify individual words that would increase or decrease listenability.

As raw corpus data are not suitable for statistical analyses of errors, each word in raw corpus data needs to be annotated with dictation tags. Dictation tags present the five types of dictation results: correct dictation, no-response dictation, deletion-type dictation, insertion-type dictation, and substitution-type dictation. Correct dictation is annotated with words that were dictated correctly, and no-response dictation with words for which learners dictated nothing. Deletion dictation is annotated with words for which learners missed a morpheme, such as deletion of the plural suffix "–s" from "governments," dictated as "government." Insertion dictation is annotated with words for which learners added a morpheme, such as the addition of an inflectional suffix "–s" to "say," dictated as "says"; or add a word, such as the addition of the article "the" to "workers," dictated as "the workers." Substitution dictation is annotated with words for which learners dictated different words, such as substituting the "d" sound for the "t" sound of "writer," dictated as "rider"; accordingly, words with substitution tags include not only the incorrect dictation but also misspelling such as "XYZ" for "written." In addition to the annotation of dictation tags, the corpus data include the listening material annotated with the part-of-speech tags in order to analyze the relation between the dictation errors and the part-of-speech.

In compiling our pilot learner corpus, each learner listened to a news report, sentence-by-sentence, and wrote down what they listened to. They were allowed to listen to each sentence three times, and were prohibited from using dictionaries or other materials during the dictation task. The hand-written transcriptions were collected, and the dictation error types were annotated on each word. The listening material was parsed with Tree Tagger [7] for annotating the part-of-speech tags.

3.4   Analysis of the Pilot Listening Learner Corpus

The present pilot learner corpus was analyzed to verify its corpus design by examining whether a language resource following such a design is suitable for the development of a listenability measurement method. As mentioned in Sect. 1, the validity of the present corpus design was assessed on the basis of extreme data distribution and reflection of well-acknowledged listenability properties. Listenability properties have been observed not only at the phonetic-phonological level, but also at the morpho-syntactic levels. As dictation tags are annotated on words, but not on phonemes, this study presents morpho-syntactic level analyses.

First, for extreme data distribution, an extremely high frequency of no-response tags was possible. Since news reports were used as the listening material, and dictation of news reports might be too difficult for beginner- and intermediate-level EFL learners, the learners may not write anything, resulting in a no-response tag. Even though words annotated with no-response tags can be analyzed as low-listenability, the no-response data fail to demonstrate how the listenability decreased, while the other error tags demonstrate the particular reason for errors. For instance, deletion-type errors show which element(s) of a word was failed in listening.

Conversely, as another instance of extreme data distribution, an extremely high frequency of correct tags was also possible. The news report taken from the VOA was linguistically controlled in consideration of language learners with respect to the vocabulary and speech rate. If this controlled news report was too easy for the EFL learners, they might write down all of the words correctly. While this is not as crucial a problem as a high frequency of no-response tags, if data include only correct tags, only analysis of what caused an increase in listenability, not a decrease, is possible.

Although 21 learners participated in this experiment, data from one learner were excluded due to incomplete dictation responses. Thus, the data analyzed in the present study were compiled from 20 learners.

Table 2 summarizes the present pilot learner corpus in terms of frequency and rate of dictation tags by part-of-speech; correct tags and the four types of error tags. The frequency is indicated by the number of observed tags, and the rate (in parentheses) was calculated by dividing the observed number by the total number of dictation tags. The items in Table 2 are sorted according to the ascending order to the no-response.

Extreme Data Distribution. With respect to extreme data distribution of no-response- and correct tags, no-response tags amounted to 0.22 in total, which appeared the most frequently among the error tags; and correct tags amounted to 0.54 in total, which appeared the most frequently among the dictation tags.

Upon further examination, the no-response tags appeared infrequently (less than 0.15) in "I'm," present verbs for non-3rd person, past-participle verbs, 'wh'-pronoun, comparative adjectives, coordinating conjunctions, cardinal numbers, adjectives, adverbs, present verbs for 3rd person, and singular nouns. Since the words annotated with these part-of-speech tags demonstrated a low rate of no-response tags, our corpus design was partially validated. In principle, the number of no-response tags should be repressed because none of the no-response tags demonstrated how learners listened to a word or a part of a word. On the other hand, while the high rate of correct tags suggests the ease of the listening material, our corpus design was validated due to the absence of extreme data distribution.

Reflection of Listenability Properties. The validity of the corpus data was first examined with respect to the well-acknowledged listenability property of whether listenability decreases for longer sentences [8]. The listenability of a sentence was calculated in terms of the mean rate of correct tags for words in the sentence, with the sentence length determined by counting the number of words in a sentence, and the correlation between listenability and the length of sentence being analyzed. This analysis confirmed a statistically significantly correlation ($r = -0.61$, $p < 0.01$). Figure 1 presents the distribution of the listenability of a sentence along the length of a sentence. In the plot, a black dot shows the listenability of the sentence 2, or the initial sentence of a text following a text title, which suggests another influence on the listenability, that is, the position of sentences in a text.

Secondly, another listenability property, whether the corpus data demonstrated low listenability due to the presence of a nominal suffix such as a plural suffix "–s," was examined. Since they are not clearly and distinctively pronounced, learners often fail to hear suffixes. The rates of correct tags were compared between singular nouns (0.60) and plural nouns (0.36), as shown in Table 2. A Wilcoxon Rank-Sum Test indicated that the rates of correct tags in singular nouns was significantly higher than that of plural nouns ($Z = 1999.0$, $p < 0.01$). In plural nouns, deletion tags (0.28) appeared more frequently than other error tags, which resulted from failure in the dictation of plural suffixes in "universities" and "governments."

Lastly, the listenability property concerning verbal suffixes, such as the past marker "–ed" or the present marker for singular third-person "–s," and whether the corpus data demonstrated errors regarding these verbal inflectional suffixes was examined. Second language learning research [9] reported that learners often incorrectly add a verbal inflectional suffix to a verb located in a syntactic context where such suffixes cannot be used. Since this type of verb has none of these suffixes, overuse of a verbal inflectional suffix results in an insertion-type error. The frequencies of error tags were compared for

Table 2. Frequency and rate of dictation tags by part-of-speech

Part-of-speech	Correct	Deletion	Insertion	Substitution	No-response
"I'm"	15 (0.75)	0 (0.00)	0 (0.00)	5 (0.25)	0 (0.00)
Verb, pres., non-3rd p.	107 (0.76)	0 (0.00)	18 (0.13)	10 (0.07)	5 (0.04)
Verb, past-participle	7 (0.35)	5 (0.25)	0 (0.00)	7 (0.35)	1 (0.05)
'Wh'-pronoun	34 (0.85)	0 (0.00)	0 (0.00)	3 (0.08)	3 (0.08)
Adj., comparative	58 (0.73)	6 (0.08)	4 (0.05)	8 (0.10)	8 (0.10)
Coordinating conjunction	73 (0.73)	0 (0.00)	0 (0.00)	16 (0.16)	11 (0.11)
Cardinal number	99 (0.62)	4 (0.03)	0 (0.00)	39 (0.24)	18 (0.11)
Adjective	249 (0.69)	0 (0.00)	6 (0.02)	64 (0.18)	41 (0.11)
Adverb	68 (0.85)	0 (0.00)	0 (0.00)	3 (0.04)	9 (0.11)
Verb, pres., 3rd p.	122 (0.61)	27 (0.14)	0 (0.00)	26 (0.13)	25 (0.13)
Noun, singular	531 (0.60)	19 (0.02)	26 (0.03)	181 (0.21)	123 (0.14)
Noun plural	231 (0.36)	176 (0.28)	7 (0.01)	119 (0.19)	111 (0.17)
'Be,' present, 3rd person	46 (0.77)	0 (0.00)	1 (0.02)	3 (0.05)	10 (0.17)
'Have,' pres., 3rd p.	17 (0.43)	0 (0.00)	0 (0.00)	16 (0.40)	7 (0.18)
Proper noun, sing.	209 (0.51)	2 (0.00)	0 (0.00)	131 (0.32)	78 (0.19)
Adv., comparative	11 (0.55)	0 (0.00)	0 (0.00)	5 (0.25)	4 (0.20)
Proper noun, pl.	18 (0.45)	12 (0.30)	0 (0.00)	1 (0.03)	9 (0.23)
Modal	50 (0.63)	0 (0.00)	1 (0.01)	10 (0.13)	19 (0.24)
Verb, gerund/participle	67 (0.30)	9 (0.04)	1 (0.00)	90 (0.41)	53 (0.24)
Verb, base form	75 (0.42)	0 (0.00)	4 (0.02)	53 (0.29)	48 (0.27)
Determiner	340 (0.61)	0 (0.00)	0 (0.00)	52 (0.09)	168 (0.30)
Preposition/subordinating conj.	322 (0.52)	0 (0.00)	0 (0.00)	71 (0.11)	227 (0.37)
Personal pronoun	42 (0.53)	0 (0.00)	0 (0.00)	7 (0.09)	31 (0.39)
'Be,' past, participle	17 (0.43)	0 (0.00)	0 (0.00)	6 (0.15)	17 (0.43)
'Be,' pres., non-3rd p.	34 (0.43)	1 (0.01)	0 (0.00)	9 (0.11)	36 (0.45)
'Wh'-determiner	15 (0.38)	0 (0.00)	0 (0.00)	7 (0.18)	18 (0.45)
Complementizer	19 (0.32)	0 (0.00)	0 (0.00)	12 (0.20)	29 (0.48)
"To"	31 (0.39)	0 (0.00)	0 (0.00)	2 (0.03)	47 (0.59)
Total	2834 (0.54)	261 (0.05)	68 (0.01)	940 (0.18)	1145 (0.22)

non-3rd person verbs in the present form (verb, present, non-3rd p.) in Table 2. Table 3 shows the observed frequency of each type of error tags, and the expected values calculated assuming equal probability. A Chi-square (1×4) test indicated the statistical significant differences among them, χ^2 (3, N = 33) = 21.4, $p < 0.01$. Ryan's multiple comparison analysis subsequently showed that insertion-type errors appeared more frequently than deletion-type and no-response-type errors ($p < 0.01$).

Summary and Limitations. Because the number of no-response tags should be further repressed, and well-acknowledged listenability properties were not exhaustively examined, the present learner corpus design was only partially validated.

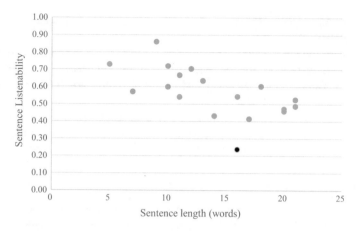

Fig. 1. Scatter plot showing sentence listenability and sentence length

Table 3. Observed frequencies and expected ratios of error tags

	Deletion	Insertion	Substitution	No-response
Observed frequency	0	18	10	5
Expected value	8.25	8.25	8.25	8.25

4 Listenability Measurement Method

4.1 Goal

In Sect. 3.4, we confirmed that our learner corpus demonstrated the sentence-length effect on listenability. Here, we further examine the validity of our corpus by developing a listenability measurement method. If our corpus design is valid, a listenability measurement method based on our corpus is expected to demonstrate acceptable performance.

4.2 Procedure

A listenability measurement method was developed with a regression analysis using our learner corpus as a language resource.

The dependent variables consisted of the correct rates of the 19 sentences. In order to calculate the correct rate of a sentence, a learner's correct rate was calculated by dividing the number of words correctly dictated by the number of words in a sentence. Then, the correct rate as the mean rate was calculated from the sum of all the learners' values of the correct rates divided by the number of learners.

The independent variables were linguistic properties of sentences, which were considered to affect listenability, and included sentence length (the total number of words in a sentence), average word length (the mean number of characters of words in a sentence), and average word levels (the mean score of word difficulty as determined by Someya [10]).

Before carrying out the multiple regression analysis, the independent variables were examined with respect to the presence of multiple-collinearity by calculating the variance of inflation factor (VIF) [11], and a multiple-collinearity of more than 10 was not found $(1.04 < \text{VIF} < 1.05)$.

Our listenability measurement method was also examined with the leave-one-out cross validation test. In this cross validation test, our method was examined N-times $(N = 19)$ by taking one instance as test data and N–1 instances as training data.

4.3   Result and Discussion

The linear combination of linguistic features was significantly related to the correct rates, F $(3, 15) = 7.21$, $p < 0.01$. The sample multiple correlation coefficient adjusted for the degrees of freedom was 0.71, and the percentage of the variance explained by the linear combination of the linguistic features was 59.1%.

The standardized partial regression coefficients are summarized in Table 4. Among the linguistic features, the average word level was not statistically significant, while the other two features were significant.

Table 4. Regression analysis result

Variable	Unstandardized coefficients	Standardized coefficients	t	p
Sentence length	−0.02	−0.54	−3.18	0.01
Average word length	−0.09	−0.47	−2.80	0.01
Average word level	−0.01	−0.01	−0.08	0.94
Intercept	1.27			

Although listenability should be explained not only with the lexical and syntactic properties but also with the phonological properties, such as phonological modification [8, 12], the present method used only the lexical (word length and word difficulty) and syntactic (the sentence length) properties as the independent variables. Despite the non-identical conditions of the independent variables, the present method was considered to demonstrate relatively high performance.

The leave-one-out cross validation test yielded a total mean squared error of 0.01, and a squared correlation coefficient of 0.59. This correlation was high, though the training data included only 19 samples. Therefore, by preparing a larger sample of training data, the performance of the listenability measurement method is expected to improve. Measurement errors from the cross validation test are shown in Table 5. Measurement error was calculated as an absolute value of the difference between an observed value and a predicted value. Our method showed more measurement error instances in the range of less than or equal to 0.05.

Table 5. Distributino of measurement error

Measurement error	≤ 0.05	≤ 0.10	≤ 0.15	≤ 0.20	≤ 0.25	≤ 0.30
Number of instances	8	7	1	0	2	1

On the basis of the high correlation and the small measurement error, our listenability measurement method was considered to demonstrate acceptable performance in the cross validation test. It was further considered that these results showed the possibility of improvement by solving the non-ideal conditions with respect to the size of training data and the lack of phonological properties.

5 Conclusion

Described is the design of a listening learner corpus for EFL learners as a language resource for a CALL system. The corpus design was developed with respect to the participants, materials, and procedures. Following this corpus design, a pilot learner corpus was compiled, and its validity as a language resource for a listenability measurement method was examined. This examination demonstrated the partial validity of the present corpus design.

Future studies to expand the corpus data, not only in size (increased number of participants and materials) but also in content (advanced-level EFL learners), develop a listenability measurement method, and implement such a tool on a CALL system, are warranted.

Acknowledgements. This work was supported in part by Grants-in-Aid for Scientific Research (B) (22300299) and (15H02940).

References

1. Beatty, K.: Teaching and Researching Computer-Assisted Language Learning. Pearson Education, London (2010)
2. Hubbard, P.: Learner training for effective use of CALL. In: Fotos, S., Browne, C. (eds.) New Perspectives in CALL for Second Language Classrooms, pp. 45–67. Lawrence Eribaum, Mahwah, NJ. (2004)
3. Petrides, J.R.: Attitudes and motivation and their impact on the performance of young english as a foreign language learners. Journal of Language and Learning **5**(1), 1–20 (2006)
4. Kotani, K., Yoshimi, T., Nanjo, H., Isahara, H.: Compiling learner corpus data of linguistic output and language processing in speaking, listening, writing, and reading. In: Wang, H., Yarowsky, D. (eds.) Proceedings of the 5th International Joint Conference on Natural Language Processing, pp. 1418–1422 (2011)
5. Rytting, C.A., Silbert, N.H., Madgavkar, M.: ArCADE: an arabic corpus of auditory dictionary errors. In: Tetreault, J., Burstein, J., Leacock, C. (eds.) Proceedings of the Ninth Workshop on Innovative Use of NLP for Building Educational Applications, pp. 109–115 (2014)

6. Ten Bosch, L., Giezenaar, G., Boves, L., Ernestus, M.: Modeling Language-learners' Errors in Understanding Casual Speech. Paper presented at Errors by Humans and Machines in Multimedia, Multimodal and Multilingual Data Processing (2015)
7. Schmid, H.: Probabilistic part-of-speech tagging using decision trees. In: Jones, D. (ed.) Proceedings of International Conference on New Methods in Language Processing, pp. 44–49 (1994)
8. Messerklinger, J.: Listenability. Center Engl. Lang. Educ. J **14**, 56–70 (2006)
9. Paradis, J.: Grammatical Morphology in Children Learning English as a Second Language: Implications of Similarities with Specific Language Impairment. Lang. Speech Hear. Sch **36**, 172–187 (2005)
10. Someya, Y.: Word Level Checker, http://www.someya-net.com/wlc/index.html
11. Neter, J., Kutner, M.H., Nachtsheim, C.J., Wasserman, W.: Applied Linear Regression Models. McGraw Hll Education, Chicago (1996)
12. Kotani, K., Ueda, S., Yoshimi, T., Nanjo, H.: A listenability measuring method for an adaptive computer-assisted language learning and teaching system. In: Aroonmanakun, W., Boonkwan, P., Supnithi, T. (eds.) Proceedings of the 28th Pacific Asia Conference on Language, Information and Computing, pp. 387–394 (2014)

The System Validation Documentation Using Text Segmentation Technique Based on CMMI Standards

Benya Painlert and Mahasak Ketcham[(⊠)]

Department Management Information Systems, Faculty of Information
Technology, King Mongkut's University of Technology North Bangkok,
Bangkok, Thailand
jjubu777@gmail.com, mahasak.k@it.kmutnb.ac.th

Abstract. This research will be presented with the principles of natural language processing, analysis and Inspection of documents based on CMMI standards. The process into a scanned PDF file used to Text Segmentation and natural language processing technique. The inspection document quality control of work efficiency is validated as compliant based on CMMI standards on Naming Convention Document by configuration manager baseline and determine with inspection manual for functional specification document and inspection manual for training document. When the determine document is sent out to work as a standard and acceptable.

Keywords: Capability maturity model integration · Natural language processing · Quality control · Text segmentation

1 Introduction

Currently CMMI version 1.3 (October 2010) [1] comprising 22 Process Areas measured by Capability or Maturity Levels by CMMI was developed by the Software Engineering Institute (SEI) [2]. Before the CMMI is used to measure the effectiveness of the procedures work as SW-CMM, SECM, IPD-CMM [3, 4] but the problem complexity some are called different names but the same was confusing. It is included in the CMMI is an important component in the optimization of production, including software is people, Procedure/Method and Tools. Which at present even CMMI is created Template up to work, but the work the also used by people all work, whether it is checking the accuracy of the application by the Test Case and the authenticity of the documents, so error occurred. Sometimes customers Reject document to revise or receive NC (Non-Compliance) from QA (Quality Assurance), which affects the reliability of the organization. So who do monitor the quality of the Product, it is very important. In order to reduce the problems caused by Human Error and increase the success of software development even more. The researchers then had the idea to develop a monitoring system for document quality control process in order to achieve CMMI work more systematically. Add the chance to succeed in the workplace, it more. Moreover, the process of quality control (QC) is one of the processes in CMMI

© Springer International Publishing AG 2018
T. Theeramunkong et al. (eds.), *Advances in Natural Language Processing,
Intelligent Informatics and Smart Technology*, Advances in Intelligent Systems

standard which is very significant examination for product and document before deliver to the client, managing the risk of software development [5]. The reputation of the organization and the Product is delivered to the customer, quality, and also creates opportunities to get a job from customers also increased.

2 Theories

Natural Language Processing: NLP is the science of this branch of science that involves many sciences, including psychology, computer engineers and statistics that will focus on the computer. Studies on the subject of natural language processing (NLP) instead of a matter of knowledge (Knowledge Representation) the techniques of parsing sentences and so on. Benefits from the computer to communicate with humans with human language itself is self-evident because it would result in the use of computers, as well as more convenient. The computer will be able to help in various areas related to language machine translation was more like a human. From one language into another language (Machine Translation) can help detect and remove computer. Analyze documents that are relevant to the matter. The computer can be used to assist in information retrieval (Information Retrieval) according to the user's computer or to a conclusion. Substantial and important issues (Information Summarization) that appeared in the papers and so on. Natural language system is a system that has been referred to the principle of academic intelligence. And knowledge over the linguistic knowledge was stored on a computerized knowledge base. The system will run a knowledge-based interpretation of knowledge and interaction with the natural language. Natural language system is a software system that features the following significant [6, 7].

2.1 The system's input and output in order to be in contact with computers. Features a natural language

2.2 Processing system to use the basic knowledge about grammar. The definition and understanding natural language.

   2.2.1 Natural Language Entity

   - Alphabet is a symbol used to represent sounds. The bulk of the restricted group.
   - Word is made up of groups of letters into words.
   - Sentence is a group of words which, when arranged together to represent meaning. With a message

   2.2.2 Natural Language Understanding

   - Morphological Analysis is an analysis of the One of the words that can be further subdivided into what I can tell that this word has nothing.
   - Syntactic Analysis is a grammatical analysis. The purpose was to see whether the sentence he received. Which consist of many word. Syntax is the structure which is the subject of the verb or object or phrase.
   - Semantic Analysis is a semantic analysis by analyzing the structure, syntax, and then determines the value of each word means anything.

- Discourse Integration is the meaning of a sentence by sentence side. Since some of the words in a sentence to understand their meanings. See the previous sentence or sentences are correct as well.

The theory of Natural Language Processing (NLP) can make the words. To be analyzed according to CMMI Level 3 standards and can effectively work as well as build confidence among customers as well.

3 Propose Method

The document used in the validation was PDF file scanned into the system to examine. The methods are following Fig. 1.

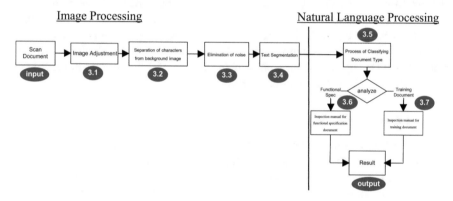

Fig. 1. Conceptual framework

3.1 Image Adjustment [8]

In this method, the adjustment was done by raising contrast to make letter part separate from the background better. The image had poor contrast because colors of the image were too bunched in the one part. If the colors were distributed over every level of the colors, the image would have better contrast. Generally, contrast adjustment is done by distributing histogram to both ends to make image look brighter and have more dimension. On the contrary, this method's expectation was not the image which was still dark becoming darker because the background of this part could be too close to the character part. Therefore, the low import value of end was set to zero, and the high value of the end was done by cutting some values to increase the range to the histogram distribution. Cut value was the part of brightness value which did not affect the pattern of characters.

3.2 Separation of Texts from Background Image

Analyzing document image, it was informed that color values in less darkness part always conformed to the color values of background which were little darker, and some values were the same values although the color values of character part were separated from the background around. Accordingly, color values could not be directly used in character pattern classification; however, the color values changed in the junction of character pattern and background could be used instead.

In this method, the experiment was held by 4 techniques which were global thresholding using one value for overall image, edge detection, contour following, and local thresholding. To compare each technique and find which technique showed better result was to validate a group of pixels which were connected component, and count the number of the objects.

3.3   Elimination of Noise

An image from thresholding was the image of black character pattern in white background. However, MATLAB program set the background to the black color, and the object is white. Therefore, the image color was reversed and edited by eliminating junction in horizontal with the size of 1 pixel (integrating bwmorph Function with hbreak Parameter). The image part near the border was set and eliminated (imclearborder Function Used), and then noise with the size of 1 pixel was eliminated (integrating bwmorph Function with clean Parameter).

3.4   Text Segmentation [9]

In the image of final result, when the difference of color value was only one level, the acquired circumference was the segmentation of the characters. The text segmentation by this process led to the knowing of coordinate value of this circumference that meant knowing the position of segmented characters. However, the value from tracking the edge of the object will be back in the matrix form which must be recorded in MATLAB's special form because the record in image form cannot be done (Figs. 2 and 3).

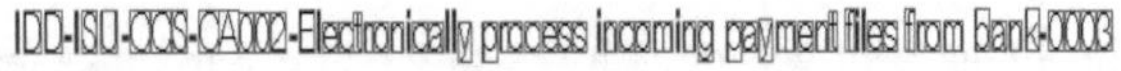

IDD-ISU-CCS-CA002-Electronically process incoming payment files from bank-0003

Fig. 2. Text segmentation

3.5   Process of Classifying Document Type

The process was to check file name which was the same name in the footer of page. The page was inputted to examine its file name whether file name setting met the regulation of document called Naming Convention in CMMI level 3. The examination was done by technique in Natural Language Processing. The processes were the following.

3.5.1   Morphology

Morphology is analysis of symbols such as document IDD-ISU-CCS-CA002-Electronically process incoming payment files from bank-0003. After analyzing, the

Fig. 3. Focus on the bottom of page for further verification

symbol gotten from the file name analysis was "–". That symbol is the separation of words in a sentence. Explanation in Fig. 4.

IDD⊟ISU⊟CCS⊟CA002⊟Electronically process incoming payment files from bank⊟0003

Fig. 4. Morphology

3.5.2   Syntactic Analysis

There is a phrase of file name after analyzing the symbols. The phrase of file name brought about structure verification which was grammatical analysis according to Naming Convention in CMMI Level 3. The aim of the analysis was to analyze inputted sentence which was composed of many words, and was analyze what the structure of the arrangement of the words was such as document IDD-ISU-CCS-CA002-Electronically process incoming payment files from bank-0003. From the first phrase, it was analyzed that the word structure from this file name was document IDD, the document in the type of Functional Specification. If the first phrase was the word "TS", the second phrase should be brought to compare for finding what type of Technical Specification of Function Specification document was, and then there was the analysis in the next process. Explanation in Fig. 5.

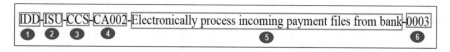

Fig. 5. Syntactic analysis

3.5.3 Semantic Analysis

Identified type of document by regulation of Naming Convention document in CMMI Level 3 such as document IDD-ISU-CCS-CA002-Electronically process incoming payment files from bank-0003. From the analysis, this separation of the word from file name was to separate the word by the result of symbols which were IDD, ISU, CCS, CA002, Electronically process incoming payment files from bank, 0003. Moreover, the structure of Function Specification document is **<Document type>** - <Work Group> - <Module> - <Sub module, Running number 3 digits> - <Process Name> - <Running number 4 digits>. Explanation Document type include in Fig. 6.

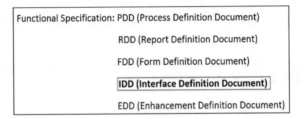

Fig. 6. Semantic analysis

3.5.4 Discourse

This process was to analyze whether the meaning of each word was in the type of structure which was specified in Naming Convention document of CMMI Level 3 such as document IDD-ISU-CCS-CA002-Electronically process incoming payment files from bank-0003. From the analysis of the meaning of the word in this file name, the meaning analysis and the structural discrimination of the word were the following:

- Document type = IDD
- Work Group = ISU
- Module = CCS
- Sub module + Running number 3 digits = CA + 002
- Process Name = Electronically process incoming payment files from bank
- Running number 4 digits = 0003

3.5.5 Pragmatic

To consider overall statement was to translate the meaning of new sentence again. Repeated translation was to find what the actual meaning based on the writer's intention. From the sample document which was IDD-ISU-CCS-CA002-Electronically process incoming payment files from bank-0003 it was summarized that the document was Functional Spec, and naming form of file name was correct to the condition specified in document of Naming Convention of CMMI Level 3 (Table 1).

When analyzing and classifying the type of document, it was recommended that the document was examined with Inspection manual for functional specification document, and Inspection manual for Training Document defined by Configuration Manager for

Table 1. Relationship of work group, module and sub module

Work group	Module	Sub module
BO	LO	PUR , MM
	FMS	BUD , AP , GL CCA , PCA , CM , AA , PA , CO
	HRM	OM , RC , PA , PD , TE , CP , BN , TM , PY
ISU	CCS	CS , BIL , DM , CA , CAI , AR , EDM
OP	MTN	PM
	WMS	WM , PS

making documents having the same standard, according to the principal of CMMI Level 3.

3.6   Inspection Manual for Functional Specification Document

The types of Function Specification document are 10 types: PDD (Process Definition Document), RDD (Report Definition Document), FDD (Form Definition Document), IDD (Interface Definition Document), EDD (Enhancement Definition Document), TS-PDD (Technical Specification Process Definition Document), TS-RDD (Technical Specification Report Definition Document), TS-FDD (Technical Specification Form Definition Document), TS-IDD (Technical Specification Interface Definition Document), and TS-EDD (Technical Specification Enhancement Definition Document).

3.6.1   The Rules of Naming Functional Specification, Technical Specification, and Training Document (BPP: Business Process Procedure)
<Document type> - <Work Group> - <Module> - <Sub module> - <Running number 3 digits> - <Process Name> - <Running number 4 digits>.

Example   naming   document   Functional   Specification   are IDD-ISU-CCS-CA002-Electronically process incoming payment files from bank-0003.

3.6.2   Version of Document
The word "Version" at the front cover should be set in English, and be followed with "Running number 4 digits" which was the same as the file name such as "Version 0001".

3.6.3   Naming of Functional Specification Document
Naming document of Functional Specification There are 4 types of identification documents will be as follows.

- EDD = "Enhancement Definition Document (EDD)" OR "ข้อกำหนดด้านการพัฒนาความสามารถของระบบงานเพิ่มเติม"
- IDD = "Interface Definition Document (IDD)" OR "ข้อกำหนดการพัฒนาการเชื่อมระบบ"
- FDD = "Form Definition Document (FDD)" OR "ข้อกำหนดด้านแบบฟอร์ม"
- RDD = "Report Definition Document (RDD)" OR "ข้อกำหนดด้านรายงาน"

3.7   Inspection Manual for Training Document

The types of Training document are 2 types: TN (Training Document), and BPP (Business Process Procedure). These types were classified by separating the first phrase which was from application of Natural Language Processing technique in topic number 3.5

3.7.1   The Rules of Naming Training Document (TN: Training Document)

<Document   type> - <Work   Group> - <Sub   Module> - <Program Name> - <Document Name> - <Running number 4 digits>

- Example document of Training document (TN) by rule 3.7.1: TN-ISU-CS-CSR045-Business partner checklist-0001
- Example document of Business Process Procedure (BPP) by rule 3.6.1: BPP-ISU-CCS-CA606-ZCAHF002- Debit Note Credit Note-0001

3.7.2   Version of Document

The word "Version" at the front cover should be set in English, and be followed with "Running number 4 digits" which was the same as the file name such as "Version 0001".

3.7.3   Naming of Training Document

The name should be named following these regulations which are

- The first line was "คู่มือปฏิบัติงาน"
- The second line was "Program Name" and
- The third line was the explanation of program name which had to relate to Functional spec of the same program.

4   Result of Experiment

The experiment of document analysis system should be correctly done according to CMMI Level 3. The experiment was evaluated from the accuracy of document examination which was: **%Accuracy = 100 − %Error**

where

$$\%\text{Error} = \text{Relative error} \times 100$$

$$\text{Relative error} = \left| \frac{x_{mea} - x_t}{x_t} \right|$$

where

Xmea   is measure value

Xt   is true value

Table 2. Results of examination papers

No.	Document	Result	Remark
1	IDD-ISU-CCS-CA006-inbound interface bank detail of customer-0005	Pass	
2	FDD-ISU-CCS-CS005-contract form-0002	Pass	
3	FDD-ISU-CCS-CAI006-แบบฟอร์มหนังสือแสดงรายละเอียดขอผลตอบแทนคืนในระกัน-0004	Pass	
4	RDD-ISU-CCS-CAN030-รายงานผู้ใช้ไฟฟ้าขอรายงานประเภทการจำแนกคืน-0002	Pass	
5	FDD-ISU-CCS-CAK001-แบบฟอร์มหนังสือแจ้งเตือนหนี้ค้างชำระ-0001	Pass	
6	EDD-ISU-CCS-CAH013-โปรแกรมหลังจากผลประโยชน์เงินประกันกับค่าไฟฟ้า-0005	Pass	
7	BPP-ISU-CCS-CA606-CAHF002 ใบพิมพ์ ใบลดหนี้-0001	Pass	
8	TS-EDD-ISU-CCS-CAD009-โปรแกรมกรุ๊ปรวมในแจ้งหนี้ค่าไฟฟ้าและในแจ้งหนี้ค่าจัดการราพนังงาน-0001	Pass	
9	RDD-ISU-CCS-BIL011-รายงานสถิติการใช้ไฟฟ้าของผู้ใช้ไฟฟ้าแต่ละราย-0002	Pass	
10	TN-ISU-CA-CACI043-download text file for E-one payment data-0001	False	Program name have symbol
11	EDD-ISU-CCS-CAC007-โปรแกรมจัดหนี้ค้างชำระ GFMIS-0002	Pass	
12	EDD-ISU-CCS-CAD001-การแจ้งหนี้แบบรวมกลุ่มบ์ Group invoicing-0004	Pass	
13	FDD-ISU-CCS-CAG001-letter inform check return-0001	Pass	
14	FDD-ISU-CCS-CS014-notification general request-0001	Pass	
15	RDD-ISU-CCS-CAN021-รายงานการชำระบาน-0003	Pass	
16	TN-ISU-CS-CSF021-ใบแจ้งหนี้ค่าจัดการราพนังงาน-0001	Pass	
17	TN-ISU-DM-DMR211-รายงานรายละเอียดประเมินหน่วยจากกับพิมต์-0001	Pass	
18	EDD-BO-LO-PUR014-validate cost element in PR-PO-0002	False	Program name have symbol
19	BPP-ISU-CCS-CA606-CAHF002 ใบพิมพ์ ใบลดหนี้-0001	Pass	
20	IDD-ISU-CCS-CAC10-outbound configuration data-0001	Pass	

The regulations of document examination were verification of front cover document, verification of document version, verification of document file name, verification of the bottom of the page, verification of document control, verification of the number of the pages that should be relevant to the bottom of the page, and to document control, and verification of the date when the last edit was that should be relevant to the document control. In all of 20 documents of experiment result, the system was able to correctly examine and analyze 18 documents, and 2 document was examined and error rate calculated to 90%. The cause of error have to symbol in document name with analysis and wraps processing errors (Table 2).

5   Conclusions

This research aimed to present the new process of document analysis and examination by applying Image Processing technique in accordance with CMMI Level 3. From the evaluation of the accuracy in the analysis and examination of 20 documents, the conclusion of accuracy in analysis and examination of the failure was 90%. The system was not able to distinguish in case of the document name confounded by the symbols. Therefore, the analysis of document structure based on the regulation identified in Naming Convention document of CMMI Level 3 was failed. In addition, the vantage of applying Image Processing and Natural language processing in product quality control according to CMMI standard in this research processing was obvious and more systematic process. Moreover, every process had clue and evidence that was able to be proved more easily and perfectly. When the process had better quality, its opportunity for success in performance was increased.

References

1. Ben, L.: [Online]. CMMI VI3 Summing up, 10 Jan 2011. Source: http://www.benlinders.com/2011/cmmi-v1-3-summing-up/
2. Software Park Thailand: [Online]. SW Companies in Thailand (2013). Source: http://www.swpark.or.th/cmmiproject/index.php/general/73-what-cmmi/
3. Zhang, L., Shao, D.: Research on combining scrum with CMMI in small and medium organizations. In: IEEE International Conference on Computer Science and Electronics Engineering (2012)
4. Wei, Q.: Integration method and example of six sigma and CMMI for process improvement. In: IEEE International Conference on Quality, Reliability, Risk, Maintenance, and Safety Engineering (QR2MSE) (2013)
5. Morakot, C., Hoa, K.D., Aditya, G.: A CMMI-based automated risk assessment framework. In: IEEE 21st Asia-Pacific Software Engineering Conference (2014)
6. ChiangMai University: [Online]. Natural Language Understanding (2010). Source: http://archive.lib.cmu.ac.th/full/T/2553/comp0353wk_ch2.pdf/
7. Informy: [Online]. Natural Language Processing (NLP), 5 Mar 2012. Source: http://informy.tumblr.com/post/18783004442/natural-language-processing-nlp/
8.

Ehsan, N., Perwaiz, A., Arif, J., Mirza, E., Ishaque, A.: CMMI/SPICE based process improvement. In: IEEE (2010)

9. Visit Yutthaphonphinit: Palm Leaf Manuscripts's Isan Dhamma Character Segmentation with Digital Image Processing. Thesis of Computer Science Faculty of Science and Technology Thammasat University (2010)

A Study of a Thai-English Translation Comparing on Applying Phrase-Based and Hierarchical Phrase-Based Translation

Prasert Luekhong[1,2]([⊠]), Taneth Ruangrajitpakorn[3,4],
Rattasit Sukhahuta[2], and Thepchai Supnithi[4]

[1] College of Integrated Science and Technology, Rajamangala University of
Technology Lanna, Chiang Mai, Thailand
prasert@rmutl.ac.th
[2] Computer Science Department, Faculty of Science, Chiang Mai University,
Chiang Mai, Thailand
[3] Department of Computer Science, Faculty of Science and Technology,
Thammasat University, Pathumthanee, Thailand
[4] Language and Semantic Technology Laboratory, National Electronics and
Computer Technology Center, Khlong Luang, Thailand

Abstract. This work presents the comparative study between phrase-based translation (PBT) and hierarchical phrase-based translation (HPBT) to find the potential and suitability for Thai-English translation. By the experiment result, we found that HPBT slightly outpaced PBT by measuring with BLEU point. The best language model was 4-gram for Thai to English translation while 5-gram gave the best result for English to Thai translation.

Keywords: Hierarchical phrase-based translation · Phrase-based translation · Statistical machine translation · Thai based translation · Translation approach comparison

1 Introduction

Many researchers on statistical machine translation (SMT) have been conducted and that resulted in several methods and approaches. The major approaches of SMT can be categorised as a word-based approach, phrase-based approach and tree-based approach [1]. With the high demand on SMT development, various tools were developed to help on implementing SMT, such as Moses [2], Phrasal [3], Cdec [4], Joshua [5] and Jane [6]. Moses and Phrasal gain our focus since they both are an open-source software and can effectively generate all three above mentioned approaches while the other cannot. However, Moses receives more public attention over Phrasal regarding popularity since it has been applied as a baseline in several consortiums such as ACL, Coling, EMNLP, and so on. With a tool such as Moses, an SMT developer, at least, requires a parallel corpus of any language pair to conveniently implement a baseline of statistical translation.

© Springer International Publishing AG 2018
T. Theeramunkong et al. (eds.), *Advances in Natural Language Processing,*
Intelligent Informatics and Smart Technology, Advances in Intelligent Systems

Various language-pairs were trialed and applied to SMT in the past such as English-French, English-Spanish, English-German, and they eventually gained much impressive accuracy results [7] since they have sufficient and well-made data for training, for instance, The Linguistic Data Consortium (LDC) [8], The parallel corpus for statistical machine translation (Europarl) [9], The JRC-Acquis [10] and English-Persian Parallel Corpus [11]. Unfortunately for a low-resource language such as Thai, the researchers suffer from insufficient data to conduct a full-scale experiment on SMT thus the translation accuracies with any other languages are obviously low, for example, simple phrase-based SMT on English-to-Thai gained BLEU score around 13.11% [12]. Furthermore, Thai currently lacks sufficient resource on syntactic tree-bank to effort on the tree-based approach hence an SMT research on Thai is limited to word-based approach and phrase-based approach. Since phrase-based SMT has been evidentially claimed to overcome the translation result from word-based approach [1], the development of word-based SMT for Thai has been dismissed.

With the limited resource for experimenting complete Thai tree-based SMT by Moses, hierarchical phrase-based translation (HPBT) becomes more interesting since its accuracies on other language pairs are severally reported to be higher than simple phrase-based translation approach (PBT) [13]. Therefore, this raises a question that which approach is more suitable for translating a Thai-English pair.

In this work, a comparative study between Thai-English translation based on HPBT and PBT approach will be conducted to set as flagship for further researches on Thai SMT. Moreover, different surrounding words (3, 4, 5 and 6-grams) as a factor will also be studied to compare as how they are affected by a translation result.

2   Development of Thai-Based SMT

This work aims to learn compatibility to a Thai-English pair from two famous approaches of SMT, i.e. phrase-based translation (PBT) and hierarchical phrase-based translation (HPBT). We design the experiment as shown in Fig. 1.

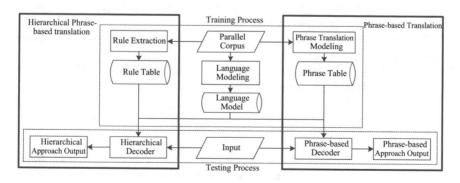

Fig. 1. A system overview

From Fig. 1, the machine translation process starts with the training process. From a parallel corpus, rules for HPBT and phrases for PBT are separately extracted to tables while the data in a parallel corpus will also use in training for generating a language model. In a summary, a training process returns three mandatory outputs for the testing process as rule table for HPBT, phrase table for PBT, and language model for both.

For testing process, input sentence for translation is needed. As the system manager input once a sentence, the input is designed to one sentence per line. To decode both approaches, each decoder executed separately and returned a translation result. For more details, each process is described in the following sections.

2.1 Phrase-Based Translation (PBT)

The statistical phrase-based MT is an improvement of the statistical word-based MT. The word-based approach uses word-to-word translation probability to translate source sentence. The phrase-based approach allows the system to divide the source sentence into segments before translating those segments. Because segmented translation pair (so called phrase translation pair) can capture a local reordering and can reduce translation alternatives, the quality of output from phrase-based approach is higher than word-based approach. It should note that phrase pairs are automatically extracted from the corpus, and they are not defined as same as traditional linguistic phrases.

As a baseline for a comparison with HPBT, PBT is developed based on 3, 4, 5, 6-gram. In this work, a phrase-based model translation proposed in [1] is implemented as follows.

1. Phrase Extraction Algorithm

The process of producing phrase translation model starts from phrase extraction algorithm. Below is an overview of phrase extraction algorithm.

(a) Collect word translation probabilities of source-to-target (forward model) and target-to-source (backward model) by using IBM model 4.
(b) Use the forward model and backward model from step (1) to align words for source sentence-to-target pair and target-to-source pair respectively. Only the highest probability is chosen for each word.
(c) Intersect both forward and backward word alignment point to get highly accurate alignment point.
(d) Fill additional alignment points using the heuristic growing procedure.
(e) Collect consistence phrase pair from step 4.

2. Phrase-Based Model

Given e is the set of possible translation results and f is the source sentence. Finding the best translation result can be done by maximize the $P(e|f)$ using Bayes's rule.

$$e^* = argmax_e P(e|f) = argmax_e(P(f|e) \times P(e)) \tag{1}$$

where $P(f|e)$ is a translation is model and $P(e)$ is the target language model. Target Language model can be trained from a monolingual corpus of the target language. Equation (1) can be written in a form of Log Linear Model to add a given customized features. For each phrase-pair, five features are introduced i.e. forward and backward phrase translation probability distribution, forward and backward lexical weight, and phrase penalty. According to these five features, Eq. (1) can be summarized as follows:

$$e^* = argmax_{e,s}P(e, s|f)$$

$$= argmax_{e,s}\left(\prod_{i=1}^{|s|}(P(\bar{f}_i|\bar{e}_i)^{\lambda_1} \times P(\bar{e}_i|\bar{f}_i)^{\lambda_2} \times P_w(\bar{f}_i, \bar{e}_i)^{\lambda_3} \times P_w(\bar{e}_i, \bar{f}_i)^{\lambda_4} \times \exp(-1)^{\lambda_5}) \times P_{LM}(e)^{\lambda_6}\right)$$

$$(2)$$

In Eq. (2) s is a phrase segmentation of f. The terms $P(\bar{f}_i|\bar{e}_i)^{\lambda_1}$ and $P(\bar{e}_i|\bar{f}_i)^{\lambda_2}$ are the phrase-level conditional probabilities for forward and backward probability distribution with feature weight λ_1 and λ_2 respectively. $P_w(\bar{f}_i, \bar{e}_i)^{\lambda_3}$ and $P_w(\bar{e}_i, \bar{f}_i)^{\lambda_4}$ are lexical weight scores for phrase pair $(\bar{e}_i, \bar{f}_i)$ with weights λ_3 and λ_4. These lexical weights for each pair are calculated from forward and backward word alignment probabilities. The term $\exp(-1)^{\lambda_5}$ is phrase penalty with feature weight λ_5. The phrase penalty score support fewer and longer phrase pairs to be selected. $P_{LM}(e)^{\lambda_6}$ is the language model with weight λ_6.

The phrase-level conditional probabilities or phrase translation probabilities can be obtained from phrase extraction process.

$$P(\bar{f}|\bar{e}) = \frac{count(\bar{f}, \bar{e})}{\sum_{\bar{f}'} count(\bar{f}, \bar{e})} \qquad (3)$$

The lexical weight is applied to check the quality of an extracted phrase pair. For a given phrase pair $(\bar{f}_i, \bar{e}_i)$ with an alignment a, lexical weight $P_w(\bar{f}_i, \bar{e}_i)$ is the joint probability of every word alignment. For a source word that aligns to more than one target word, the average probability is used.

$$P_w(\bar{f}, \bar{e}) = P(\bar{f}|\bar{e}, a) = \prod_{i=1}^{|\bar{f}|} \frac{1}{|\{j|(i,j) \in a\}|} \sum_{\forall(i,j)\in a} w(f_i|e_j) \qquad (4)$$

where $w(f_i|e_j)$ is lexical translation probability of the word pair $(f_i|e_j)$ and $|\bar{f}|$ is a number of word in phrase $\bar{f}$.

3. Decoding

The decoder is used to search the most likely translation e^* according to the source sentence, phrase translation model and the target language model. The search algorithm can be performed by beam-search [14]. The main algorithm of beam search starts from an initial hypothesis. The next hypothesis can be expanded from the initial hypothesis which is not necessary to be the next phrase segmentation of the source sentence. Words in the path of hypothesis expansion are marked. The system produces a translation alternative when a path covers all words. The scores of each alternative are calculated and the sentence with highest score is selected. Some techniques such as hypothesis recombination and heuristic pruning can be applied to overcome the exponential size of search space.

2.2 Hierarchical Phrase-Based Translation (HPBT)

Chiang [13] proposed a hierarchical phrase-based translation (HPBT) and described it as a statistical machine translation model that uses hierarchical phrases. Hierarchical phrases are defined as phrases consisting of two or more sub-phrases that hierarchically link to each other. To create hierarchical phrase model, a synchronous context-free grammar (aka. a syntax-directed transduction grammar [15]) is learned from a parallel text without any syntactic annotations. Asynchronous CFG derivation begins with a pair of linked start symbols. At each step, two linked non-terminals are rewritten using the two components of a single rule. When denoting links with boxed indices, they were re-indexed the newly introduced symbols apart from the symbols already present.

In this work, we follow the implement instruction based on Chiang [13]. The methodology can be summarized as follows.

Since a grammar in a synchronous CFG is elementary structures that rewrite rules with aligned pairs of right-hand sides, it can be defined as:

$$X \rightarrow \langle \gamma, \alpha, \sim \rangle \tag{5}$$

where X is a non-terminal, γ and α are both strings of terminals and non-terminals, and $\sim$ is a one-to-one correspondence between nonterminal occurrences in γ and non-terminal occurrences in α.

1. Rule Extraction Algorithm

   The extraction process begins with a word-aligned corpus: a set of triples $\langle f, e, \sim \rangle$, where f is a source sentence, e is an target sentence, and $\sim$ is a (many-to-many) binary relation between positions of f and positions of e. The word alignments are obtained by running GIZA++ [16] on the corpus in both directions, and forming the union of the two sets of word alignments. Each word-aligned sentence from the two sets of word alignments is extracted into a pair of a set of rules that are consistent with the word alignments. This can be listed in two main steps.

   (a) Identity initial phrase pairs using the same criterion as most phrase-based systems [17], namely, there must be at least one word inside one phrase aligned to a word inside the other, but no word inside one phrase can align to a word outside the other phrase.
   (b) To obtain rules from the phrases, they look for phrases that contain other phrases and replace the sub-phrases with nonterminal symbols.

2. Hierarchical-phrase-based Model

Chiang [13] explained hierarchical-phrase-based model that "Given a source sentence f, a synchronous CFG will have many derivations that yield f on the source side, and therefore many possible target translations e."

With such explanation, a model over derivations D is defined to predict which translations are more likely than others. Following the log-linear model [18] over derivations D, the calculation is obtained as:

$$P(D) \propto \prod_i \varnothing_i(D)^{\lambda_i} \tag{6}$$

where the $\varnothing_i$ are features defined on derivations and the λ_i are feature weights. One of the features is an m-gram language model $PLM(e)$; the remainder of the features will define as products of functions on the rules used in a derivation:

$$\varnothing_i(D) = \prod_{(X \to \langle \gamma, \alpha \rangle) \in D} \theta_i(X \to \langle \gamma, \alpha \rangle) \tag{7}$$

Thus we can rewrite $P(D)$ as

$$P(D) \propto PLM(e)^{\lambda_{LM}} \times \prod_{i \neq LM} \prod_{(X \to \langle \gamma, \alpha \rangle) \in D} \theta_i(X \to \langle \gamma, \alpha \rangle)^{\lambda_i} \tag{8}$$

The factors other than the language model factor can put into a particularly convenient form. A weighted synchronous CFG is a synchronous CFG together with a function w that assigns weights to rules. This function induces a weight function over derivations:

$$w(D) = \prod_{(X \to \langle \gamma, \alpha \rangle) \in D} w(X \to \langle \gamma, \alpha \rangle) \tag{9}$$

If we define

$$w(X \to \langle \gamma, \alpha \rangle) = \prod_{i \neq LM} \theta_i(X \to \langle \gamma, \alpha \rangle)^{\lambda_i} \tag{10}$$

then the probability model becomes

$$P(D) \propto PLM(e)^{\lambda_{LM}} \times w(D) \tag{11}$$

3. Training

On the attempt to estimate the parameters of the phrase translation and lexical-weighting features, frequencies of phrases are necessary for the extracted rules.

For each sentence pair in the training data, more than one derivation of the sentence pair use the several rules extracted from it. They are following Och and others, to use heuristics to hypothesize a distribution of possible rules as though then observed them in the training data, a distribution that does not necessarily maximize the likelihood of the training data. Och's method [17] gives a count of one to each extracted phrase pair occurrence. They give a count of one to each initial phrase pair occurrence and then distribute its weight equally among the rules obtained by subtracting sub-phrases from it. Treating this distribution data, They use relative-frequency estimation to obtain $P(\gamma|\alpha)$ and $P(\alpha|\gamma)$. Finally, the parameters λ_i of the log-linear model Eq. (11) are learned by minimum-error-rate training [19], which tries to set the parameters so as to maximize the BLEU score [20] of a development set. This gives a weighted synchronous CFG according to Eq. (6) that is ready to be used by the decoder.

4. Decoding

We applied CKY parser as a decoder. We also exploited beam search in the post-process for mapping source and target derivation. Given a source sentence, it finds the target yield of the single best derivation that has source yield:

$$\hat{e} = e\left(\text{argmax}_{D\,s.t\,f(D)}\,P(D)\right) \quad (12)$$

They find not only the best derivation for a source sentence but also a list of the k-best derivations. These k-best derivations are utilized for minimum-error-rate training to rescore a language model, and they also use to reduce searching space by cube pruning [21].

3 Experiment

This work aims to compare the quality of Thai-English SMT between phrase-based translation (PBT) approach and hierarchical phrase-based translation (HPBT) approach hence we set the experiment as follow.

3.1 Data Preparation

The bilingual corpus used in this experiment contains Thai-English parallel sentences gathered from two sources: BTEC (Basic Travel Expression Corpus) [22] and HIT London Olympic Corpus [23]. The total number of the sentence pairs is 150,000 sentences while 1000 sentences were selected off the pool by humans as a development set for tuning weights in the translation process. Thai sentences were automatically word-segmented. Punctuations, such as full-stop and question mark, were removed from both languages.

3.2 Experiment Setting

A language modeling tool, which is SRILM [24], was exploited to generate a language model of 3, 4, 5, and 6-gram. To prevent the bias results from the sample data, ten-fold cross-validation was applied in the experiment. BLEU score [20] was used as a translation evaluator. The higher score of BLEU indicates the more likeliness of a translation result to a reference sentence.

3.3 Results

We evaluate the system from both directions, Thai-to-English and English-to-Thai. The evaluation involves in 3, 4, 5 and 6-gram from PBT and HPBT. Figures 2 and 3 show

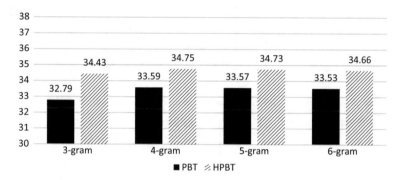

Fig. 2. An average BLUE score of Thai to english translation based on different n-grams and approaches

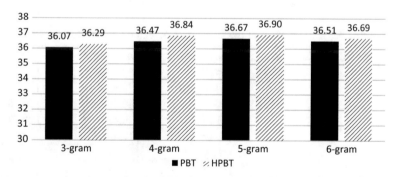

Fig. 3. An average BLUE score of english to Thai translation based on different n-grams and approaches

the translation results in term of BLEU point.

From the result, a 4-gram model of HPBT provided the best translation result for Thai-to-English and it gained 1.12 bleu points over PBT approach. We hence look into details in each sentence translation. We separated the translation results into three parts that are low, mid and high bleu point. We then observe the change in bleu points of HPBT and split them into five ranges as very poor (HPBT $\leq$ 0.5 bleu point to PBT), poor (HPBT < 0.5 and < 0 bleu to PBT), same (HPBT = PBT in bleu), good

(HPBT > 0 and > 0.5 bleu points to PBT) and very good (HPBT $\geq$ 0.5 bleu point to PBT). From these settings, we obtain Tables 1, 2 and 3.

Table 1. A comparison of translation results with 0–0.33 BLEU points

Sentence with low BLEU point (0–0.33)	Comparison range	Amount of sentence	Percentage
	Very poor	–	–
	Poor	14,503	17.71
	Same	46,395	56.66
	Good	19,941	24.35
	Very good	1043	1.27
	SUM	81,882	100

Table 2. A comparison of translation results with 0.34–0.66 BLEU points

Sentence with medium BLEU point (0.34–0.66)	Comparison range	Amount of sentence	Percentage
	Very poor	38	0.10
	Poor	8411	21.69
	Same	24,781	63.89
	Good	4746	12.24
	Very good	809	2.08
	SUM	38,785	100

Table 3. A comparison of translation results with 0.67–1.0 BLEU points

Sentence with high BLEU point (0.67–1.0)	Comparison range	Amount of sentence	Percentage
	Very poor	2038	5.40
	Poor	4065	10.78
	Same	30,470	80.77
	Good	1150	3.05
	Very good	–	0
	SUM	37,723	100

From the result in tables, we found that HPBT performs better for the sentences with low BLEU point since it has better results in low BLEU set of sentence. In the medium BLEU point set, HPBT and PBT mostly produced the same result for 63.89%. However, there are roughly 20% that HPBT yielded a worse score. We investigated into this issue and found that 95% of the cases are the short input sentences in range of 3–4 words. We also found that 3093 sentences were exactly translated as same as a reference sentence (1.0 BLEU points) while PBT translated to lower BLEU score (>0 and <1.0), and there are 217 sentences that HPBT gained 1.0 BLEU point while PBT obtain 0.0 point.

4 Conclusion

In this work, we studied by applying 3, 4, 5 and 6-gram PBT model and HPBT model to translate Thai-to-English and English-to-Thai. By comparing the results, we found that HPBT showed potential on translating of Thai-English translation in 10 fold cross-validation experiment. That results shown HPBT get higher translation result than PBT for each language model. Moreover, the suitable language model is 4-gram for Thai to English translation while 5-gram is the best for English to Thai translation for both approaches.

Acknowledgements. This work was supported by the Office of the Higher Education Commission, Thailand for supporting the grant fund under the program Strategic Scholarships for Frontier Research Network for the Ph.D. Program.

References

1. Koehn, P.: Statistical Machine Translation. Cambridge University Press (2010)
2. Koehn, P., Hoang, H., Birch, A., Callison-Burch, C., Federico, M., Bertoldi, N., Cowan, B., Shen, W., Moran, C., Zens, R., Bertoldi, N., Cowan, B., Shen, W., Moran, C., Zens, R., Bertoldi, N., Cowan, B., Moran, C., Dyer, C., Constantin, A., Herbst, E., Hoang, H., Birch, A., Moses: Open source toolkit for statistical machine translation. In: Proceedings of the 45th Annual Meeting of the ACL on Interactive Poster and Demonstration Sessions, June, pp. 177–180 (2007)
3. Cer, D., Galley, M., Jurafsky, D.: Phrasal: a toolkit for statistical machine translation with facilities for extraction and incorporation of arbitrary model features. In Proceedings of the NAACL, June, pp. 9–12 (2010)
4. Dyer, C., Weese, J., Setiawan, H., Lopez, A.: Cdec: a decoder, alignment, and learning framework for finite-state and context-free translation models. In: Proceedings of the ACL, July, pp. 7–12 (2010)
5. Schwartz, L., Thornton, W., Weese, J.: Joshua: an open source toolkit for parsing-based machine translation. In: Mach. Translation, March, pp. 135–139 (2009)
6. Vilar, D., Stein, D., Huck, M.: Jane: open source hierarchical translation, extended with reordering and lexicon models. In: On Statistical Machine Translation and Metrics MATR (WMT 2010), July, pp. 262–270 (2010)
7. Matrix Euro. [Online]. Available: http://matrix.statmt.org/. Accessed 29 May 2012
8. Liberman, M., Cieri, C.: The creation, distribution and use of linguistic data: the case of the linguistic data consortium. In: Proceedings of the 1st International Conference on Language Resources and Evaluation (LREC) (1998)
9. Koehn, P.: Europarl: a parallel corpus for statistical machine translation. In: MT Summit, **11** (2005)
10. Steinberger, R., Pouliquen, B., Widiger, A.: The JRC-Acquis: a multilingual aligned parallel corpus with 20 + languages. Arxiv Prepr. cs/0609058. **4**(1) (2006)
11. Yazdchi, M.V., Faili, H.: Generating english-persian parallel corpus using an automatic anchor finding sentence aligner. In: Natural Language Processing and Knowledge Engineering (NLP-KE), International Conference on 2010 , pp. 1–6 (2010)
12. Porkaew, P., Ruangrajitpakorn, T.: Translation of noun phrase from english to thai using phrase-based SMT with CCG reordering rules. In: Pacling (2009)

13. Chiang, D.: Hierarchical phrase-based translation. Comput. Linguist. **33**(2), 201–228 (2007)
14. Koehn, P.: Pharaoh: a beam search decoder for phrase-based statistical machine translation models. In: Machine Translation From Real Users to Resolution, pp. 115–124 (2004)
15. II, P.L., Stearns, R.: Syntax-directed transduction. J. ACM, **I**(3), 465–488 (1968)
16. Och, F.J., Ney, H.: Giza++: Training of Statistical Translation Models. Internal report, RWTH Aachen University, http://www.i6.informatik.rwth-aachen.de (2000)
17. Och, F.J., Ney, H.: The alignment template approach to statistical machine translation. Comput. Linguist. **30**(4), 417–449 (2004)
18. Och, F.J., Ney, H.: Discriminative training and maximum entropy models for statistical machine translation. In: Proceedings of the 40th Annual Meeting on Association for Computational Linguistics, July, pp. 295–302 (2002)
19. F.J. Och, Minimum error rate training in statistical machine translation. In: 41st Annual Meeting on Association for Computational Linguistics (ACL), vol. 1001, no. 1, pp. 160–167 (2003)
20. Papineni, K., Roukos, S., Ward, T., Zhu, W.: BLEU : a method for automatic evaluation of machine translation. Comput. Linguist. 311–318 (2002)
21. Feng, Y., Mi, H., Liu, Y., Liu, Q.: An efficient shift-reduce decoding algorithm for phrased-based machine translation. In: Proceedings of the 23rd International Conference on Computational Linguistics: Posters, pp. 285–293 (2010)
22. BTEC Task International Workshop on Spoken Language Translation. [Online]. Available: http://iwslt2010.fbk.eu/node/32. Accessed 27 May 2012
23. Yang, M., Jiang, H., Zhao, T.: Construct trilingual parallel corpus on demand. Chinese Spok. Lang. Process. 760–767 (2006)
24. Stolcke, A.: SRILM-an extensible language modeling toolkit. In: Seventh International Conference on Spoken Language Processing (2002)

English-Cebuano Parallel Language Resource for Statistical Machine Translation System

Zarah Lou B. Tabaranza, Lucelle L. Bureros, and Robert R. Roxas[✉]

Department of Computer Science, University of the Philippines Cebu, Cebu, Philippines
{zarahloutabaranza, ohmworx3}@gmail.com,
robert.roxas@up.edu.ph

Abstract. This paper describes the building of an English-Cebuano parallel corpus to be fed into a Statistical Machine Translation System. This parallel corpus is built as a combination of human translation and automatic translation. The automatic translation involves the use of web crawlers that automatically identify bilingual websites that contain English and Cebuano web pages in them. The extracted pages underwent language filtering to exclude those that are not written in English or Cebuano. The extracted and identified pages underwent preprocessing to clean and structure the texts. Then a sentence-level text alignment algorithm was employed to align sentences from an English text to a Cebuano text. The result of the sentence-level alignment and the output of the human translators comprised the parallel English-Cebuano corpus that will eventually be fed into a Statistical Machine Translation System.

Keywords: English-Cebuano parallel corpus · Manual translation · Automatic translation · Parallel text mining · Language filtering · Text alignment

1 Introduction

Efforts to come up with a good Machine Translation System for natural languages started several decades already but have not yet stopped. It is because it is a very interesting but difficult endeavor. Several techniques were employed and new techniques are still being explored because natural languages are very dynamic and continue to evolve.

Some believe that practical machine translation must be constructed from a heuristic point of view rather than from a pure rigid analytical linguistic method [10]. Some believe that machine translation systems must employ the context-driven approach [1, 14, 16, 17]. But the most common approach used is the statistical approach [3, 4], which heavily uses parallel corpus as the training data. This approach popularizes the Statistical Machine Translation (SMT) system [6, 8]. The accuracy of the output of any SMT systems depends on the availability of parallel corpus with high quality as its training data. Parallel corpus, as pairs of text that are translations of each other, can either be manually or automatically acquired. Manual acquisition of parallel corpus is more accurate but expensive. So it is much more ideal to build parallel corpus automatically.

© Springer International Publishing AG 2018
T. Theeramunkong et al. (eds.), *Advances in Natural Language Processing,*
Intelligent Informatics and Smart Technology, Advances in Intelligent Systems

Several parallel corpora have existed and have been primarily used as the training data for building Statistical Machine Translation Systems. Common examples are English-German Translation Corpus, English-Norwegian Parallel Corpus (ENPC), English-Swedish Parallel Corpus (ESPC), etc. [9]. World Wide Web (www) is a great source of building parallel corpus because of its size, and it covers several languages in different aspects [13]. Unfortunately, web does not directly extract a parallel corpus on specific languages because the extraction needs further organization and specification to get the appropriate information.

2   Related Works

Parallel corpus has been made to many language pairs like English-Swahili [12], English-Croatian [7], French-English [4], Chinese-English [5, 11], English-Hindi [2] just to name a few. These are for major languages in the world. Seldom can we find a parallel corpus that includes the minor languages, which are found in some local places only.

There is none yet for English-Cebuano pair. The tools and technology in the field of natural language processing continues to improve giving the researchers the opportunity to explore the impossible. Unfortunately, parallel corpus as training data for the translation services for languages that are not widely used over the web such as Cebuano has not yet been given much attention. Acquisition of texts for scarce languages from the web requires thorough language processing since technologies that processes these languages are limited. Thus an endeavor on building English-Cebuano parallel corpus will increase the awareness of researchers to include other minor languages as training data for SMT systems.

3   Methodology

In building the English-Cebuano parallel corpus, the manual and automatic approaches were employed. In the manual approach, human translators were hired to translate documents either from English to Cebuano or vice versa. In the automatic approach, web crawlers were constructed to automatically find and retrieve Web documents that were written both in English and in Cebuano. The outputs of both approaches were combined to come up with the parallel language resource to be fed into the Statistical Machine Translation System. Figure 1 shows the steps taken to build the Parallel Language Resource for English-Cebuano corpus.

3.1   The Manual Approach

In this approach, four (4) human translators were hired to do the translations. They were given source documents that were written either in English or Cebuano, and they translated them to Cebuano or English, respectively. Then some experts in both English and Cebuano languages were hired to evaluate the quality of the translations. Any of the human translators who failed to meet the standard was not allowed to continue the

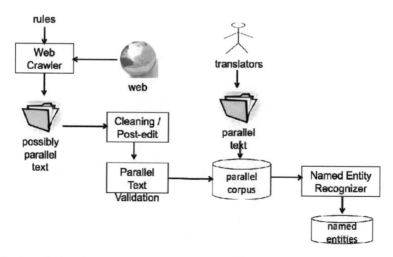

Fig. 1. Steps in building the parallel language resource for English and Cebuano languages

work, and his/her translation works were discarded as this could affect the quality of the training data for the Statistical Machine Translation System.

3.2 The Automatic Approach

Based on the existing studies, there are major phases in building a parallel corpus: Parallel Text Mining, Language Filtering, Preprocessing, and Text Alignment as illustrated in Fig. 2. These phases have to be implemented individually and eventually integrated into the system.

To get the English and Cebuano language-pair from the web, bilingual websites have to be identified and crawled from the web to yield English and Cebuano corpora. The language filtering and preprocessing must both be implemented to prepare the pairs of sentences, which are subject for text alignment. The alignment results, with English sentences and their corresponding Cebuano translations, comprise the parallel corpus.

Parallel Text Mining The list of bilingual websites was both automatically and manually identified. In the automatic identification, a function that automatically fetched website URLs from the web was implemented. With Browser from Splinter installed from pip, the system could automatically input a search query in the Google Search Engine. It was assumed that website URLs with "*lang=ceb*" contained contents in Cebuano language and "*lang=en*" contained contents in the English language. To be able to generate the English counterpart of the retrieved websites, the URLs of these websites containing "*lang=ceb*" were replaced with "*lang=en.*" The URLs with both Cebuano and English pages were retrieved and saved in a file.

In manual identification, the web was searched manually to locate bilingual websites. The Bible is one of the documents that are available on several languages, thus it is included on the list of bilingual websites. We made sure that the Cebuano version is a translation of the same version of the Bible in English because there are many versions of the Bible, and it is not wise to use different versions for the source and the target.

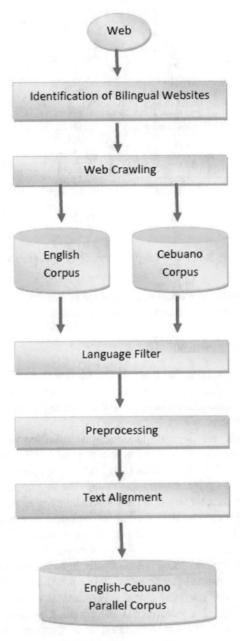

Fig. 2. The process flow

A general web crawler was created to extract the contents of the particular website included in the list of collected parallel websites by the automatic identification. The web page of each website was requested via Python requests and returned its HTML contents. It was found out that most sentences on web pages appear within the HTML *<p>* tag, which stands for paragraph. The extracted texts were saved as a raw text file on a *.txt* format and counted as a single text file. The websites collected by the manual identification had a another web crawler and parser that would take care of the specific HTML formats.

Language Filtering The language detector initially built a language model for both English and Cebuano languages. The language model is a dictionary where the key is equal to the language code such as *"en"* and *"cb,"* which both map to a list of most commonly used words in the specified language. The top 200 most commonly used Cebuano words came from http://www.binisaya.com/node/962/817, while the top 200 most commonly used English words came from http://teacherjoe.us/Vocab200.html. The language detector would accept an input text to be classified as *English*, *Cebuano*, or *None*. Built-in functions from the Natural Language Toolkit (NLTK) were used to tokenize the given texts into sentences and then each sentence into a list of tokens. Through the language detector, text files with languages that did not belong to the selected language pair were disregarded.

Using the top 200 most commonly used Cebuano words and the top 200 most commonly used English words suffices for the language detection task. It is always possible that an English document contains some foreign words, Cebuano words for instance. But these foreign words are very limited in an English document. So if the top 200 commonly used Cebu words happened to be applied to an English document to check whether or not the document is written in Cebuano, it will turn out that the document is not a Cebuano document because of the high probability that those top 200 commonly used Cebuano words are not found in that English document. But if the top 200 commonly used English words happened to be applied to an English document to check whether or not the document is English, it will turn out that the document is an English document because of the high probability that those top 200 commonly used English words are found in that English document.

In the same manner, it is always possible that a Cebuano document contains some foreign words, English words for instance. But these foreign words are very limited in a Cebuano document. So if the top 200 commonly used English words happened to be applied to a Cebuano document to check whether or not the document is English, it will turn out that the document is not an English document because of the high probability that those top 200 commonly used English words are not found in that Cebuano document. But if the top 200 commonly used Cebuano words happened to be applied to a Cebuano document to check whether or not the document is Cebuano, it will turn out that the document is a Cebuano document because of the high probability that those top 200 commonly used Cebuano words are found in that Cebuano document.

This approach will be able to detect any English or Cebuano documents of various length in spite of the fact that only the top 200 commonly used words are employed to test whether a particular document is written in Cebuano or in English.

Language filtering was only implemented or applied on the corpus collected from the automatically identified bilingual websites because such corpus was expected to be unorganized and unreliable. For the corpus taken from the manually identified bilingual websites, there was no need to apply the language detector because the language detection was done by the human being. Figure 3 illustrates how the language detector worked with a given text to be detected whether *English*, *Cebuano*, or *None*. The given text was tokenized and converted the list of the tokens into a *token_set*. A set is a Python data structure that is a collection of unique objects which would remove the duplicates of tokens in the set. The expression that computed the value of intersection resulted the union of the two sets, which was 2 (*'over'*, *'the'*) and 0 (*None*) for English and Cebuano sets, respectively. The language detector gave *English* as the language of the text since it scored larger than Cebuano. The corpora whose languages were correctly identified as *Cebuano* or *English* then underwent preprocessing.

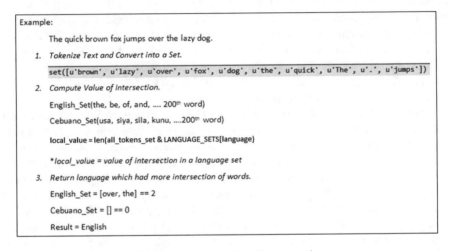

Fig. 3. A language detection example

Preprocessing The preprocessing included the process of storing the original sentences, removing punctuation marks, converting the sentences into lowercase, tokenizing, and filtering the *stop words* as illustrated in Fig. 4. The sentences have to be filtered out to get the *significant elements*. One way to identify *significant elements* was to remove *stop words* in a sentence, and the words that remained comprised the "*word_list*" of each sentence. *Stop words* are common short function words such as *the, is, at, which, on*, and others. Words not found in the list of *stop words* were considered as *significant elements*. In order to remove duplicate words in the "*word_list*" of significant elements in each sentence, the "*word_list*" was converted into a set. The "*word_list*" and the cleaned sentences were saved for the next phase, that is, text alignment.

Text Alignment Given a pair of texts, each sentence from the source text was matched with a sentence from the target text using the modified version of Sidorov et al.'s algorithm [15] as shown in Eq. 1. We modified Eq. 1 and used it in this study because when we tested such equation with some example pairs of sentences, it turned out that in many cases the result for both possible pairs is equal, when in fact, one of the pairs is really a translation of each other. If the algorithm produced the same similarity score despite the fact that one of the pairs is a translation of each other, then

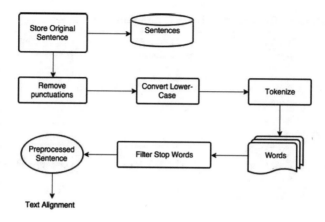

Fig. 4. Implemented methods of preprocessing

the algorithm is problematic and cannot produce the desired result. There were also instances that some pairs that were not a translation of each other had the lowest similarity score, and therefore, should be taken as the parallel text, when in fact the sentences are not translations of each other. Therefore, we modified Eq. 1 by removing the *CharLengthDiff*(e, c) part of the equation as shown in Eq. 2.

$$
\begin{aligned}
Similarity(e,c) &= DictonaryDiff(e,c) \\
&+ SignificantElementsDiff(e,c) \\
&+ CharLengthDiff(e,c)
\end{aligned}
\tag{1}
$$

$$
\begin{aligned}
Similarity(e,c) &= DictonaryDiff(e,c) \\
&+ SignificantElementsDiff(e,c)
\end{aligned}
\tag{2}
$$

A sentence match was assigned with a score that represented the similarity of the two sentences (e, c). The similarity equation is the sum of the differences of the two sentences in terms of significant elements and dictionary differences only. Smaller similarity score means high possibility that the two sentences are translations of each other.

This appears to be contradictory at first glance. But it is not because the similarity score is the sum of two differences. This means that when the sum of the differences is high, the possibility that the two sentences are similar is low. Table 1 illustrates the point, where the first pair with zero similarity score is the correct translation of each other.

DictionaryDiff(e, c) represented the number of significant elements that were not mutual translations of each other. Equation 3 shows the equation for dictionary difference of two sentences using Sidorov et al.'s algorithm [15]. Each significant element of an English sentence was looked up in an online bilingual Cebuano dictionary to check for the existence of such a significant word. If it existed in the dictionary, the translation count was incremented. The number of word pairs that matched was doubled in order to balance the equation since each translation match covered two words.

Table 1. Text alignment computation example of two sentence pairs

Sentences	Significant words	No. of translations	Dictionary difference	Significant element difference	Similarity score
Cebu is a nice destination	Cebu, nice, destination	3	3 + 3 − (2 × 3) = 0	|3 − 3| = 0	0 + 0 = 0
Nindot nga destinasyon ang Cebu	Cebu, nindot, destinasyon				
Cebu is a nice destination	Cebu, nice, destination	1	3 + 3 − (2 × 1) = 4	|3 − 3| = 0	4 + 0 = 4
Niadto ko sa Cebu kagahapon	Niadto, cebu, kagahapon				

The total number of translations was subtracted from the total number of significant elements of the two sentences to obtain the number of words in which the sentences differed.

$$DictionaryDiff(e, c) = SignificantElements(e) \\ + SignificantElements(c) - (2 * Translations(e, c)) \tag{3}$$

SignificantElementsDiff(e, c) is the absolute value of the difference of number of significant elements in two sentences. Each English sentence is associated with the Cebuano sentence that yielded the least similarity score, and the two sentences were taken as a pair for the alignment.

If a particular English-Cebuano sentence match yielded the least similarity score, but the English sentence already existed in the alignment, the system would compare which similarity score was lesser between the English-Cebuano pairs that existed in the alignment and would choose the pair with the least similarity score. This ensured that no sentences were aligned twice in the alignment. The result of the alignment was a list of pairs of English and Cebuano sentences comprising the parallel corpus.

4   Results and Discussion

Before employing the language filtering and text alignment algorithms, they were tested with some data in order to determine the accuracy rate of both algorithms. If the accuracy rate of both algorithms is high, then it is legitimate to use them as components of a system that automatically builds the English-Cebuano parallel corpus, and we can expect good results.

4.1   Language Filtering Algorithm

The experiment to evaluate the language filtering algorithm was conducted on English and Cebuano corpora. The English and Cebuano corpora used for the evaluation were the news articles from *Sun Star* and *SuperBalita* Cagayan De Oro for the year 2015. These corpora were crawled from the *Sun Star* Website Archives. The goal of the language filtering algorithm was to detect whether a given text was either *English*, *Cebuano*, or *None*. The language filtering algorithm was expected to detect the language with the highest value of intersection from a given text. There were 324 out of 365 English news articles that were correctly identified as *English*. The accuracy rate is 88.77%. There were 201 out of 202 Cebuano news articles that were correctly identified as *Cebuano*. The accuracy rate is 99.5%. If we get the average accuracy rate for both languages, that gives us an average accuracy rate of 94.14%. The incorrectly identified news articles were either identified as *Cebuano* when it was actually *English* and vice versa, or the language identifier yielded *None*.

4.2 Text Alignment Algorithm

The text alignment algorithm was tested on an English corpus with 610 sentences and in a Cebuano corpus with 432 sentences. These corpora contained parallel texts that needed to be aligned. The corpus used was the texts extracted from 3 text files taken from automatically identified bilingual websites and 6 text files from the manually identified bilingual websites. The goal of the text alignment was to align sentences in order to locate the parallel texts. If all Cebuano sentences have parallel English sentences, then there must be 432 sentences that were parallel sentences between the two corpora. The text alignment was able to identify 299 parallel sentences because not all Cebuano sentences have parallel English sentences. To validate whether these 299 identified parallel sentences were correct translations of each other, the results were manually labeled. It turned out that there were 270 parallel sentences that were correctly identified, and 29 were incorrect. Thus, the text alignment algorithm yielded 90.3% accuracy rate.

4.3 Parallel Text Mining

Both the automatic and manual approach to mining parallel text generated a list of candidate bilingual websites. The automatic approach, which fed an input query to a search engine, yielded 106 websites that actually contained "*lang=ceb*" within their URLs. These websites produced lesser text files than the expected 106 files because there were websites that were filtered out by the language filtering algorithm or did not have English counterpart. The quality of the bilingual websites from the automatic approach, however, was unreliable because the Cebuano web pages were mostly powered by Google Translate. The Cebuano translations need enormous human intervention to rectify the erroneous translations. Without human intervention, the parallel corpus will have a very low quality and will spoil the quality of the translation output that will be generated by the Statistical Machine Translation System, into which this parallel corpus will be fed.

The manual approach to identifying bilingual websites yielded a few bilingual websites and some bilingual documents. These manually identified bilingual websites were not powered by Google Translate, thus they were reliable, and we got more files from them. Table 2 shows the quantities of the collected corpus and the maximum number of sentence pairs that were parallel, which was computed by the minimum number of text files multiplied by the average sentences per text file.

Table 2. Quantitative data on collected English and Cebuano corpus quantity

Quantity	Manual collection	Automatic collection
English text files	3145	72
Cebuano text files	3153	71
Average sentences per text file	17	24
Average words per sentences	10	18
Maximum number of sentence pairs	53,465	1704

Based on the result, there were more text files produced from the manually identified bilingual websites than from the automatically identified bilingual websites. This was due to the difference in the method of extracting the website contents. In the automatically identified bilingual websites, for one URL containing "*lang=ceb*" within it, this results to one text file if it has the corresponding English website with "*lang=en*" in its URL. It does not go deeper because, if that same website has other parallel pages, they are already counted separately.

Websites generated from the manual approach to identifying bilingual websites had their own web crawlers, which were able to crawl the websites on more than one level, retrieving the website's archive. For example, one Bible website with English and Cebuano versions of the Bible has several English web pages with corresponding Cebuano web pages. These parallel web pages are very reliable because they were products of human translation, not powered by Google Translate.

4.4   Combined Parallel English-Cebuano Corpus

As for the parallel texts produced by the human translators, so far they have translated 584 English files into Cebuano and 161 Cebuano files into English with a total of 731,379 words or approximately 60,948 sentences using the average of 12 words per sentence. So combining the manual and automatic approach to translation, we have initially collected about 116,117 English-Cebuano sentence pairs that comprise the parallel English-Cebuano corpus.

5   Conclusion and Future Works

The building of English-Cebuano parallel language resource for Statistical Machine Translation System has been presented. Both human translation (manual) and parallel text mining (automatic) approaches were used. The human translation so far produced approximately 60,948 English-Cebuano parallel sentences and the parallel text mining so far produced 53,465 and 1704 English-Cebuano parallel sentences. Combining both methods, we have built a parallel English-Cebuano corpus with about 116,117 parallel sentences.

The language identifier and text alignment algorithms were tested with some data in order to determine the accuracy rate of both algorithms. The language filtering algorithm yielded an average accuracy rate of 94.14%, and the text alignment algorithm yielded an accuracy rate of 90.3%. These results justify the use of them as components of a system that automatically builds the English-Cebuano parallel corpus.

Because the project is still on-going, it is expected that the number of sentence pairs that we currently have in our English-Cebuano parallel corpus will increase greatly by the end of 2016. This corpus will eventually be fed into the Statistical Machine Translation System.

Acknowledgements. This work is funded by the Commission on Higher Education (CHED) of the Republic of the Philippines for the ASEAN-MT project.

References

1. Amarado, M.A.A.: A prototype context-driven English-to-Japanese machine translation system that translates compound and complex sentences (March 2011), special Problem, University of the Philippines Cebu
2. Aswani, N., Gaizauskas, R.: A hybrid approach to align sentences and words in english-hindi parallel corpora. In: Proceedings of the ACL Workshop on Building and Using Parallel Texts, pp. 57–64. (June 2005)
3. Brown, P., Cocke, J., Pietra, S.D., Pietra, V.D., Jelinek, F., Mercer, R., Roossin, P.: A statistical approach to french/ english translation. In: Second International Conference on Theoretical and Methodological Issues in Machine Translation of Natural Languages, (June 1998)
4. Brown, P.F., Pietra, S.A.D., Pietra, V.J.D., Mercer, R.L.: The mathematics of statistical machine translation: parameter estimation. Comput. Linguist. **19**(2), 263–311 (1993)
5. Chen, J., Nie, J.Y.: Automatic construction of parallel English-Chinese corpus for cross-language information retrieval. In: Proceedings of the Sixth conference on Applied Natural Language Processing (ANLP-NAACL-2000), pp. 21–28, (2000)
6. Déchelotte, D., Schwenk, H., Bonneau-Maynard, H., Allauzen, A., Adda, G.: A state-of-the-art statistical machine translation system based on moses. In: MT Summit XI, pp. 127–133, Sept 2007
7. Esplà-Gomis, M., Klubička, F., Ljubešić, N., Ortiz-Rojas, S., Papavassiliou, V., Prokopidis, P.: Comparing two acquisition systems for automatically building an English-Croatian parallel corpus from multilingual websites. In: Ninth International Conference on Language Resources and Evaluation (LREC 2014), pp. 1252–1258, May 2014
8. Koehn, P.: Statistical machine translation. http://www.statmt.org/
9. Nesselhauf, N.: Corpus linguistics: a practical introduction (September 2011). http://www.as.uni-heidelberg.de/personen/Nesselhauf/files/Corpus%20Linguistics%20Practical%20Introduction.pdf
10. Nitta, Y., Okajima, A., Yamano, F., Ishihara, K.: A heuristic approach to English-into-Japanese machine translation. In: Horecky, J. (ed.) Proceedings of the Ninth International Conference on Computational Linguistics, pp. 283–288. North-Holland Publishing Company, Dec 1922

11. Pauwa, G.D., Wagacha, P.W., de Schryver, G.M.: Mining Chinese-English parallel corpora from the web. In: Third International Joint Conference on Natural Language Processing (IJCNLP 2008), pp. 847–852, Jan 2008
12. Pauwa, G.D., Wagacha, P.W., de Schryver, G.M.: The sawa corpus: a parallel corpus English—Swahili. In: Proceedings of the EACL 2009 Workshop on Language Technologies for African Languages AfLaT 2009, pp. 9–16, March 2009
13. Resnik, P., Smith, N.A.: The web as a parallel corpus. Comput. Linguist. **29**(3), 349–380 (2003)
14. Roxas, R.R., Villarino, M.G.C.H.: A prototype context-driven Cebuano-English machine translation system. In: Proceedings of the PSITE-76th Regional Convention, pp. 13–16, April 2010
15. Sidorov, G., Posadas-Durán, J.P., Jiménez-Salazar, H., Chanona-Hernandez, L.: A new combined lexical and statistical based sentence level alignment algorithm for parallel texts. Int. J. Comput. Linguist. Appl. **2**(1–2), 257–263 (Jan–Dec 2011). http://www.gelbukh.com/ijcla/2011-1-2/
16. Tamayo, K.D.B.: A prototype english to waray-waray context-driven machine translation system (Apr 2012), special Problem, University of the Philippines Cebu
17. Valmores, J.D.: A prototype tagalog to english context-driven machine translation system (March 2009), special Problem, University of the Philippines Visayas Cebu College

A Part-of-Speech-Based Exploratory Text Mining of Students' Looking-Back Evaluation

Toshiro Minami[1(✉)], Sachio Hirokawa[2], Yoko Ohura[1],
and Kiyota Hashimoto[3]

[1] Kyushu Institute of Information Sciences, 6-3-1 Saifu, Dazaifu,
Fukuoka 818-0117, Japan
minami@kiis.ac.jp; minamitoshiro@gmail.com
[2] Kyushu University, Fukuoka, Japan
[3] Prince of Songkla University, Phuket, Thailand

Abstract. In our lectures at universities, we observe that the students' attitudes affects a lot to their achievements. In order to prove this observation based on data, we have been investigating to find effective methods that extract students' attitudes from lecture data; such as examination score as an index to student's achievement, attendance and homework data for his/her effort, and answer texts of the term-end questionnaire as information source of attitude. In this chapter, we take another approach to investigate the influences of words used in the answer texts of students on their achievements. We use a machine learning method called Support Vector Machine (SVM), which is a tool to create a model for classifying the given data into two groups by positive and negative training sample data. We apply SVM to the answer texts for analyzing the influences of parts of speech of words to the student's achievement. Even though adjectives and adverbs are the same in the sense that they modify nouns and verbs, we found that adverbs affects much more than adjectives, as a result. From our experiences so far, we believe that analysis of answers to the evaluations of students toward themselves and lectures are very useful source of finding the students' attitudes to learning.

Keywords: Attitude to learning · Answer texts to questionnaire · Influence of part of speech · Educational data mining · Support vector machine (SVM)

1 Introduction

Efforts to improve education and learning have been directed more to an evidence-based approach in which a variety of educational and learning data are employed to analyze the reality and to improve educational practice, evaluation, and policy, together with promoting learners' self-improvements by giving more chances for self-awareness. Educational data mining is expected to offer valuable findings with these data, and much research has been conducted [2–4, 12, 14].

In Japan, in particular, these efforts are part of Faculty Development activities, which are mandatory since 2005. Most of the Japanese universities conduct questionnaire surveys for each course, though they are usually anonymous ones and are

T. Theeramunkong et al. (eds.), *Advances in Natural Language Processing,*
Intelligent Informatics and Smart Technology, Advances in Intelligent Systems

difficult to conduct a deep analysis to find correlations with their academic performances. Thus, we have been conducting inscribed questionnaire surveys that students have to write their names in various courses, and have tried deeper analyses to find correlations with their descriptions and their academic performances [7–11].

One of the purposes of our surveys is to promote self-introspection on the learners' side, and their textual descriptions are supposed to be filled with what they think they have learned, what they felt and thought about the class, how they evaluated themselves, etc., as well as how they evaluated the class and the instructor (See the full set of questions at Table 1). Our findings so far [7–11] are summarized in Sect. 2.3, and they are based on the appearances of some seemingly suggestive content words, which is often the case with other related studies.

Table 1. Questions: (Q1)–(Q5) for class evaluation and (Q6)–(Q11) for self evaluation

(Q1)	What did you learn in this class? Did it help you?
(Q2)	What are the good points of the lectures?
(Q3)	What are the bad points that need to be improved?
(Q4)	What score you give to the lectures as a whole? (With the numbers from 0 to 100, where the pass level is 60 as in the same way to the examination score.)
(Q5)	Write comments on the course, on the lectures and the lecturer, if any,
(Q6)	What are your good points in learning attitudes and efforts for the course?
(Q7)	What are your bad points that should have to be improved?
(Q8)	How do you evaluate of your diligence and eagerness to study? Choose one of "excellent," "good," "fair," "rather poor," and "poor".
(Q9)	Have you made any questions to the lecturer? Choose one of "made questions more than once," "made questions once," "had no questions," "could not make questions," and "had no questions at all". Describe in detail about the question(s) you made, and if the lecturer answered appropriately,
(Q10)	Have you done some research or information retrieval in order to find the answers of some questions after school hours? Choose one of "retrieved often," "retrieved sometimes," "had not retrieved for solving questions," and "had no questions at all". Describe in detail about what you have done,
(Q11)	What score you give to yourself as the evaluation of your own efforts and attitude toward the course. (With the numbers from 0 to 100 as in the same way as in Q4), and finally,
(Q12)	Write other comments, if any.

In this chapter, we investigate a more formal aspect of their descriptions, parts of speech of the words appearing there, employing a machine learning method, Support Vector Machine, to find correlations between this formal aspect and their academic performances. This idea originally comes from the fact that essay and report writing by students with higher performances tend to contain effective uses of transitions and other words of particular parts of speech. Our research question in this paper is: Which parts of speech of words appearing in students' textual descriptions in our questionnaire surveys have correlations with students' academic performances and why?

The result of our analysis reveals a surprising and intriguing fact: the use of degree adverbs has a positive correlation with students' performances, which have never been mentioned in other studies. We admit that the size of our sample is small and it represents just a small part of the students in target, university students, but this correlation is rather naturally explained from a linguistic viewpoint, and this finding, as well as others, shows the usefulness of focusing on some linguistically formal aspects in educational data mining.

The rest of this chapter is organized as follows: Sect. 2 refers to related works of educational data mining in Sect. 2.1, description of the data in Sect. 2.2, and our previous work in Sect. 2.3. Section 3 describes our analytical method introducing Support Vector Machine and evaluative measurements. Section 4 describes the result and discussion, and Sect. 5 concludes this chapter.

2 Related Work

2.1 Related Works of Educational Data Mining

As more efforts to improve education have been expected from the society, particularly since late 1990s in Japan, more students' questionnaire surveys have been conducted on courses and schools. Their main purposes were to find more about what students thought and felt about teachers, courses, and schools, and to reflect those results to faculty evaluations, and thus most of them were conducted anonymously and thus it has been impossible to compare those results with students' academic performance, though some universities introduced e-portfolio and give feedbacks to each students. On the other hand, some teachers voluntarily conducted questionnaire surveys for their courses [5], often called looking-back ones, whose purposes include giving students opportunities to check what they understand and how they learn. Our survey is categorized as the latter type.

For the usual anonymous surveys, there have been many statistical studies [1, 6], most of which were published in technical or research reports of each university. Some studies employ statistical methods from correlational analysis to PCA to more advanced methods, and they tried to find what teachers should improve and how students were motivated. The relations between the survey replies and academic performances have also pursued in many studies [16, 17], and they found that students' academic performance has a positive correlation with their activeness, involvedness, and other self-motivated minds found in questionnaire surveys. In other words, their results strongly implied that their survey replies are correlated to their active mind towards study and, as a result, their survey results were correlated to their academic performances. However, most of them employed categorical data but not text data [13], and text analysis of students' questionnaire surveys are more needed to complement those results and to find new suggestions on educational and learning improvements.

2.2   The Target Data

The target data were generated in the class named "Information Retrieval Exercise" for 2009 [7–11]. The number of students attending the class was 36.

The major aim of the course is to let the students become expert information searchers so that they have sufficient knowledge about information retrieval, search, finding, and also have enough skills in finding appropriate search engine site and search keywords based on the understanding of the aim and background of the retrieval.

The course consists of 15 lectures. The attendant is supposed to solve a couple of small quizzes at every lecture, which turns into the attendance record and the basic data for attendance score. Homework is additionally assigned every week in every lecture. It aims to give an opportunity to the students of reviewing what they have learned in the class, and to study preliminary knowledge for the next class. At the same time, the students are requested to write a lecture note every week, which also aims to make the students review what they have learned.

The term-end examination of the course consists of 3 problems/questions. The first question is about finding the Web sites of search engine, and summarizing their characteristic features, together with discussing about the methods for information retrieval. The second question is on finding the Web sites on e-books and on-line material services. The third question is to find and discuss about the information criminals in the Internet environment.

These questions aim to evaluate the skill on information retrieval including the planning and summarizing skills that are supposed to be learned and trained in the course as they do their exercises in the classes and as they do their homeworks. The scores of term-end examination represent the evaluation results according to this aim.

We also imposed a term-end looking-back questionnaire asking the students for looking back what they have learned, what they did, etc., and evaluate themselves and the class. Table 1 shows the list of questions.

The aim of the questionnaire is basically to provide the students with the opportunity for the Check and Act steps of the PDCA (Plan, Do, Check, and Act) cycle for the lecture. It is also a good source for the instructor to know about the students and to improve the lecture next time.

2.3   Our Previous Findings

In this subsection, we illustrate findings we had in our previous studies according to the article [11]. We have found that the examination scores and the word usage in the answer to the question (Q1) were correlated. For example, the students in the highest score group used such words that indicate their interest in the world affairs and comparing in and out of Japan; such as "Foreign," "Overseas," "Japan," and "National." On the other hand, the students in the lowest score group used the words such as "Remember," "Learn," "Useful"; thus they looked like good students superficially. However, by considering their poor scores, it is highly possible that they took attention to the lecture-related terms themselves alone, presumably without understanding what they really mean and how they relate one another.

In order to confirm our observation, we assigned one of the four types to each word/expression used in the answer texts. The 4 types are as follows: (A) Those which directly relate to the lecture's main topic of "information retrieval"; e.g., library, retrieval method, key/keyword, library's opening/closing time, PC, etc., (B) Those which relate to the subjects which are taken in the lectures, but are not directly related to the main topic; e.g., IC tag system (for libraries), introduction (of libraries where the lecturer has visited), user-operated check-out machine, etc., (C) General words which appear frequently in the answers of students; e.g., know, go, way/method, think, etc. Note that the frequently used words are those which appear more than twice; about 20% of general words. Thus the students who use high amount of words in this type, will attend the class and learn in the same way as many other students, and (D) General words which do not appear frequently: e.g., use, inner part, tackle with, route, object, etc. Thus the student who uses a lot of words in this type, might have her own point of view, which is different from those of other students. In this sense, she has some kind of originality.

We estimated student's score by summing up all the estimated grades for the types from A to D. Take some students for example. Student st003's score changes from 99 to 74, st014, from 90 to 73, and st030, from 27 to 62.2.

Figure 1 shows the correlation between estimated and original scores. We can see that the general rule also holds here because the gradient of the approximation line is positive (0.31). So we may roughly evaluate the students' scores according to the word types they used. We need to collect more data in order to evaluate even more accurately. The graph shows that the students having maximum or nearly maximum estimated scores are located in the upper-middle group, i.e., from near the average score 66.4 up to near 80, in their original scores. With one exception of st007, all the students in this region take the estimated scores more than 60. On the other hand, the top two students (st014 and st003) in the original scores have much lower scores in the estimated ones.

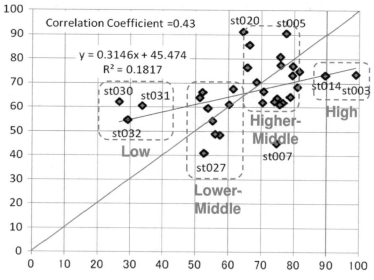

Fig. 1. Correlation of original/real score (x-axis) and estimated score (y-axis)

Now we can rephrase our findings in a more precise statement: The students having very high scores may not take attention to the wide areas of topics. The students in the high-middle group pay attention not only to the directly-related topics of the course, but also to other topics as well. The students having the lower-middle scores prefer to study the topics directly-related topics of the course. The students having very low scores may not be very bad in their attitudes. Their poor scores may come from different reasons.

3   Method: Achievement-Word Correlation Analysis with Support Vector Machine

Support Vector Machine (SVM) is a machine learning algorithm which classifies the data into two groups; one as positive and the other as negative. A typical clustering process using SVM consists of two steps; (1) model creation, and (2) classification of data. In step (1), sample vector data are given for training. The vector data are given to SVM with flags which indicate if the vector is positive sample or negative one. After training, SVM creates a model. Then in step (2), we put a test vector to the model, and the model returns whether the test vector is judged as positive or negative.

In our approach, we utilize the values of the model as the weight of the component as in a similar method taken in [15]: a component corresponds to a word in our case. As the weight shows the distance of the word from the maximum-margin hyperplane, which separate the positive and negative samples. Thus the sign of the weight of word indicates whether the word has effects which side to belong to, and its absolute value indicates the amount of positiveness or negativeness.

Now we apply this method to our target data. Figure 2 shows the distribution of the examination scores in the decreasing order. The first student has the maximum examination score (nearly 100). On the other hand, the last student (35th one) has the minimum score about 25. Median is 70 in our case.

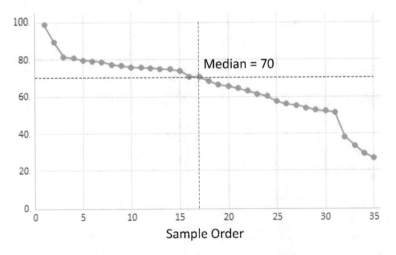

Fig. 2. Distribution of examination scores in the decreasing order

Then we set the learning problem as to classify the students into two groups with same size (i.e., 17 each); one for those who have scores >70, or high group, and those having <70, or low group. The sample vector for a student is an array of values 1 or 0 for the word which appears in the answer texts to either of the questions. Thus, if the value is 1 for a word w, it means that the student uses the word w at least once in some of the answers of the student. This word-vector is given to SVM as a positive sample if the student have the examination score >70, and it is given as negative if <70. As the result SVM created a model; a word-vector which consists of the weight of each word.

For our experiment, let us modify the word-vector described in the previous paragraph. As we would like to analyze considering the parts of speech, we add up tagged words to the vector. We use the tags "n" for noun, "v" for verb, "adj" for adjective, and "adv" for adverb. For example if a word w is a noun, we add the component corresponding to "$n{:}w$."

In order to extract the weighting values of a word w, we make a vector which has value 1 only in the component for the word w, and apply the model to this vector. In this way, we can get the weight of the word w, or $weight(w)$. This is a measure for the importance of a word, which we call "w.o" measure for importance of words. The absolute value $|weight(w)|$ of the weight can be used as a measure for importance as well, which we call "w.a".

Next, we try to find a group of words that are most effective in predicting the examination scores. Table 2 shows the prediction performance based on the top N positive and bottom N negative words with respect to the measure w.a.

In this example, we chose top $2N$ words in terms of the $w.a$ measure for importance of words. The first line ($N = 0$) shows the baseline case where all the words are used for classification. Let us define the numbers tp, fp, tn, and fn as follows: tp (true positive) is the number of students who are classified to high group in the model, and are actually in high group, fp (false positive) is the number of students who are classified to high group in the model, but are actually in low group. tn (true negative) is the number of students who are classified to low group in the model, and are actually in low group, fn (false negative) is the number of students who are classified to low group in the model, but are actually in high group.

Then the exactness measures are defined as follows:

$$\text{Precision} = \frac{tp}{tp+fp}$$

$$\text{Recall} = \frac{tp}{tp+fn}$$

$$\text{F-measure} = 2\frac{\text{Recall} \cdot \text{Precision}}{\text{Recall} + \text{Precision}}$$

$$\text{Accuracy} = \frac{tp+tn}{tp+fp+tn+fn}$$

Table 2. Prediction performance with 2 N words based on the w.a measure

N	Precision	Recall	F-measure	Accuracy
0	0.4845	1.0000	0.6424	0.4845
1	0.7267	0.7767	0.7314	0.7345
2	0.5762	0.8667	0.6886	0.6512
3	0.6262	1.0000	0.7648	0.7179
4	0.6010	0.8333	0.6906	0.6798
5	0.6333	0.9333	0.7394	0.7048
6	0.6367	0.9500	0.7509	0.7131
7	0.7000	1.0000	0.8180	0.7964
8	0.7467	1.0000	0.8390	0.8131
9	0.6262	1.0000	0.7624	0.7012
10	0.6262	1.0000	0.7624	0.7012
20	0.7500	1.0000	0.8466	0.8298
30	0.7200	1.0000	0.8252	0.7881
40	0.7767	1.0000	0.8644	0.8548
50	0.7867	1.0000	0.8652	0.8381
60	0.8667	1.0000	0.9152	0.9048
70	0.7367	1.0000	0.8366	0.8131
80	0.6967	1.0000	0.8144	0.7881
90	0.6967	1.0000	0.8144	0.7881
100	0.6667	1.0000	0.7929	0.7631
200	0.5295	0.8333	0.6389	0.5845
300	0.4845	1.0000	0.6424	0.4845

From the table, we can see that the best prediction performance can be reached at $N = 60$ with F-measure $= 0.9152$ and Accuracy $= 0.9048$, both of which are better than the baseline ($N = 0$) that uses all words.

4   Result and Discussion

Table 3 shows the top 20 words having highest weights in terms of $w.a$. The 4th column "df" and the 5th column "tf" show the document frequency and the term frequency of a term, respectively. The 6th column "sum-tf" shows the summation of the term frequencies "tf" at and above the line.

The adverbs "adv:とても (very)," "adv:あまり (not very)," and "adv:もう少し (a little more)" appear as the 2nd, 5th, and 20th words, respectively. Nouns "n:宿題 (homework)" and "n:私 (I)" also appear as the high-ranked words; 9th and 11th, respectively.

As we see how the word "adv:とても (very)" appears in the sentences, it is used in order to positively evaluate the lecture:

Table 3. Characteristic words in terms of *w.a* (top 20)

Rank	Weight*df	Weight	df	tf	sum-tf	Word
1	−6.2868	−0.2028	31	128	128	n:検索 (search)
2	2.8365	0.1891	15	26	154	adv:とても (very)
3	0.8065	0.1613	5	7	161	n:調べ (investigate)
4	−2.0644	−0.1588	13	21	182	n:仕方 (method)
5	1.3662	0.1518	9	14	196	adv:あまり (not very)
6	2.5488	0.1416	18	18	214	n:普通 (normal)
7	3.6736	0.1312	28	121	335	v:できる (can)
8	−0.5108	−0.1277	4	5	340	n:種類 (variety)
9	2.9279	0.1273	23	88	428	n:宿題 (homework)
10	−2.3446	−0.1234	19	48	476	n:情報 (information)
11	1.5756	0.1212	13	18	494	n:私 (I)
12	0.3390	0.1130	3	4	498	n:際 (when)
13	−0.4324	−0.1081	4	4	502	v:増やす (increase)
14	0.5380	0.1076	5	7	509	v:いける (can)
15	−0.7490	−0.1070	7	7	516	n:たくさん (a lot of)
16	−0.4236	−0.1059	4	4	520	v:忘れる (forget)
17	0.9423	0.1047	9	19	539	v:聞く (hear)
18	−1.1352	−0.1032	11	12	551	v:つける (attach)
19	2.5450	0.1018	25	54	605	v:調べる (investigate)
20	−0.3021	−0.1007	3	4	609	adv:もう少し (a little more)

- とても勉強になりました (very good to learn)
- とても楽しい授業でした (very enjoyable lectures)
- とても面白かった (very much interesting)
- とても役立つ (very useful)
- とても分かりやすい説明 (very easy to understand)
- とても参考になりました (very informative)

On the other hand, the word "adv:あまり (not very)" is used in order to make negative evaluations on the student herself with looking herself back:

- あまり調べることもなかった (did not investigate not very much)
- あまり PC と縁のない私 (I did not use PC very much)
- あまり得意ではなかった (not very much good at)
- タイピングがあまり速くない (typing is not very fast)
- あまり上手く検索できなかった (could not search very well)
- あまり工夫できなかった (could not devise very much)

Table 4 shows the number of occurrences and the probability $Pr(p)$ for each part-of-speech "p."

Table 4. Global probability of part-of-speech

Part-of-speech	All	Noun (n)	Verb (v)	Adjective (adj)	Adverb (adv)
Occurrences	5200	2923	1709	277	256
Probability	1.00	0.56	0.33	0.05	0.05

Figure 3 shows the ratio $Pr(p, N)$ of the top N words for each part-of-speech "p," i.e., $Pr(p, N)/Pr(p)$, where $Pr(p, N)$ is the probability of occurrence of the words with the part-of-speech"p" in the top N words measured by $w.a$. For example, there are two adverbs within top 5 rank. "adv:very" appears at the 2nd rank with 26 occurrences and "adv:not very" appars at the 5th rank with 14 occurrences. Thus, there are 40 occurrences of adverbs in top 5 rank and $Pr(\text{adv}, 5) = 40/196 = 0.2041$. On the other hand, we have $Pr(\text{adv}) = 256/5200 = 0.0492$ from Table 4. Therefore, we have $Pr(\text{adv}, 5)/Pr(\text{adv}) = 0.2041/0.0492 = 4.1483$, which means that adverbs occur 4.15 times more frequently within top 5 rank compared to the average.

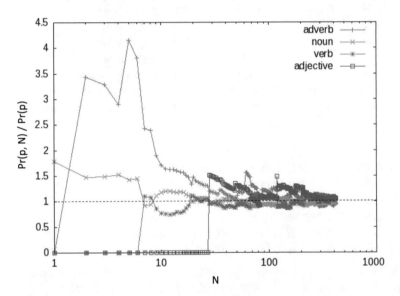

Fig. 3. Probability comparison of parts-of-speech in top K feature words

The results give us an interesting suggestion. Some degree adverbs like "adv:とても (very)," "adv:あまり (not very)," and "adv:たくさん (a lot of)" are ranked high, as we mentioned above. The first two show their positive correlation to the students' performance while the third has a negative correlation.

As we already saw, "adv:とても (very)" often appears in a positive context in terms of class/teacher while "adv:あまり (not very)" often appears in a negative context in terms of students themselves. As this survey had replies signed by students, we admit that students tend to write more positive impressions towards class and teacher under such a condition, but what this result suggests is that both "adv:とても (very)" and "adv:あまり (not very)" are positively correlated to their academic performances because these words, as a characteristics of degree adverb, mostly co-occur with evaluation phrases: those students who write more evaluative sentences, whether they are about classes or about themselves, are more likely to win higher academic performances.

The reason why "adv:たくさん (a lot of)" is negatively correlated may also be explained: "adv:たくさん (a lot of)" is an indefinite quantifier of big numbers that are often used when the writer is not sure of the exact number and he/she actually does not understand it well.

From the viewpoint of text mining, this suggestion is useful. Researchers usually try to extract meaningful evaluation words and others, and meet the difficulty that those words does not occur so frequently as expected. Our result suggests that, instead of focusing on evaluation words themselves, focusing on degree adverbs which are expected to modify evaluation words also tells part of students' attitudes that are related to their academic performances.

5   Concluding Remarks

In this chapter, we investigated some relations between the examination scores of students and the words used in their answer texts at the term-end self and lecture evaluation questionnaire. In this study, we investigated differences among the parts of speech of words, and have found that adverbs are more correlated to the examination score than other parts of speech. On the other hand, adjectives are found to be little correlated to the examination score.

These results may have a variety of explanations. One possibility is that the students in the high scored group recognize things using verbs, in other words "dynamic, active learning," and thus they use adverbs as modifiers, whereas those in the low scored group are inclined to use nouns, or "static, passive learning" and they use adjectives more than average. Another possibility is that the frequent use of degree adverbs come from the frequent use of evaluation phrases and that those who try to make more evaluations tend to be more self-aware, which is expected to contribute to their study.

By investigating the attitudes of students from more than one points of view, we are able to have a more sophisticated understanding of students as learners. We expect that such an approach will eventually open up the new methods for providing us with useful tips on how to improve the motivation of students.

References

1. Arimichi, Y.: Analysis of a survey on english classes. Res. Rep. Takamatsu Nat. Coll. Technol. **44**, 9–36 (2009). (in Japanese)
2. Baker, R.S.J.D., Yacef, K.: The state of educational data mining in 2009: a review and future visions. J. Educ. Data Min. Article 1, **1**(1), (2009)

3. Bienkowski, M., Feng, M., Means, B.: Enhancing Teaching and Learning Through Educational Data Mining and Learning Analytics—An Issue Brief. U.S. Department of Education, Office of Educational Technology (2012)
4. El-Halees, A.M.: Mining Students Data to Analyze Learning Behavior: A Case Study. ResearchGate (2009)
5. Goda, K., Hirokawa, S., Mine, T.: Correlation of grade prediction performance and validity of self-evaluation comments. ACM SIGITE **2013**, 35–42 (2013)
6. Ishida, T., Kumoi, G., Goto, M., Goto, Y., Hirasawa, S.: An evaluation model for e-learning based on student questionnaires. In: JASMIN 10, p. 4, (2010) (in Japanese)
7. Minami, T., Ohura, Y.: Lecture data analysis towards to know how the students' attitudes affect to their evaluations. In: 8th International Conference on Information Technology and Applications (ICITA 2013) (2013)
8. Minami, T., Ohura, Y.: Investigation of students' attitudes to lectures with tex-analysis of questionnaires. In: 4th International Conference on E-Service and Knowledge Management (ESKM 2013), p. 7, (2013)
9. Minami, T., Ohura, Y.: A correlation analysis of student's attitude and outcome of lectures—investigation of keywords in class-evaluation questionnaire. In: DTA 2014 as a part of the 6th International Mega-Conference on Future Generation Information Technology (FGIT 2014) (2014)
10. Minami, T., Ohura, Y.: Towards improving students' attitudes to lectures and getting higher grades–with analyzing the usage of keywords in class-evaluation questionnaires. In: Seventh International Conference on Information, Process, and Knowledge Management (eKNOW 2015) (2015)
11. Minami, T., Ohura, Y.: How student's attitude influences on learning achievement? An analysis of attitude-representing words appearing in looking-back evaluation texts. Int. J. Database Theory Appl. **8**(2), 129–144 (2015)
12. Romero, C., Ventura, S.: Educational data mining: a survey from 1995 to 2005. Expert Syst. Appl. **33**, 135–146 (2007)
13. Romero, C., Ventura, S., Espejo, P., Hervas, C.: Data mining algorithms to classify students. In: 1st International Conference on Educational Data Mining (EDM 2008), pp. 8–17, (2008)
14. Romero, C., Ventura, S.: Educational data mining: a Review of the State of the Art. IEEE Trans. Syst. Man Cybern. Part C Appl. Rev. **40**(6), 601–618 (2007)
15. Sakai, T., Hirokawa, S.: Feature words that classify problem sentence in scientific article. In: IIWAS 2012, pp. 360–367, (2012)
16. Shigaki, I.: The relationship between scores of student questionnaires to teaching and examination scores. Mem. Osaka Inst. Technol. Ser. A **55–1**, 1–9 (2010). (in Japanese)
17. Yatsufusa, T., Wang, R., Satonobu, J., Ishii, Y.: Correlation analysis between questionnaires on lectures and GPA in Hiroshima Institute of Technology. Bull. Hiroshima Inst. Technol. (Education) **14**, 69–74 (2015). (in Japanese)

Finding Key Terms Representing Events from Thai Twitter

Apivadee Piyatumrong[✉], Chatchawal Sangkeettrakarn,
Choochart Haruechaiyasak, and Alisa Kongthon

National Electronics and Computer Technology Center (NECTEC), 112
Thailand Science Park, Phahonyothin Rd., Khlong Nueng, Khlong Luang,
Pathumthani 12120, Thailand
{apivadee.piy, chatchawal.san, choochart.har, alisa.kon}
@nectec.or.th

Abstract. In the fast and big data era, we all desire to understand trend or big picture of a story instantly. This work wants to find an automatic approach to extract the good-enough key terms of each event appear in Thai Twitter society. The core idea is to help reducing time for human to do the key term extraction, yet the quality of such selected key terms are acceptable by human and is better than our previous implementation. Our studied approaches focus to work on Thai language and covered preprocessing, feature selections and weighting schemes on three Thai real tweet events with different characteristics. Our experiment comprise four main approaches and a number of hypothesis. Our findings confirm the usefulness of hashtag terms with five or more character length, the benefit of bigram with stop words and the importance of event characteristics. In fact, we conclude to use different approaches for different types of event. The performance and rational evaluations are done by statistical analysis, evaluators voting, and F-Score measurement and are confirmed to be better than previous work twice as much.

1 Introduction

Nowadays, Twitter is used for propagating information in real-time i.e. news headline, crisis outbreak, stock market trend, traffic situation, etc. Moreover, it serves people to be able to update personal interested topics such as k-pop idol related info or television show, etc. It also has become a marketing tool, not only for big company or celebrities, but for not-a-celebrity people to be able to speak their voices and to offer their services broadly. An insight understanding over the information flow over Twitter can bring forth the power to utilize this particular media, both as a communication mean to reach target group and/or as the human sensor to feed back information of interest. Thus, company as well as the not-a-celebrity people desires to experience such abilities as soon as possible. For that to happen, fundamental needs are to grab the most understandable and relevance keywords explaining the event-of-interest. Then these users can act according to the characteristics about the event, or to what people mostly give information about the event. Our goal is to implement a better process for extracting 20 key features of interested event from Thai tweet. In order to accomplish the goal, we face two big challenges. Firstly, working with Thai tweet has challenges of processing the evolving Thai language over social media. This challenge is much accelerating as the social media spreading broader into Thai society. New terms and slang are recreating everyday and being adopt widely. Secondly, the previous implemented operation is hugely depending on human. The system administrator has to select seed words, for crawling relevance tweets, then select key terms for representing the event. This is an expensive operation that we want to overcome with an automated process to suggest key features.

The article is organized as follows, Sect. 2 explains basic structure of tweet and challenges found with Thai language tweet. Section 3 gives an overview of prior works relating to keyword extraction and Thai text tokenization. Followed by our proposed approaches found in Sect. 4. Then Experiment Settings, Sect. 5, explains the 3 tweet events data set used in this work, the implementation using Apache Spark and validation method by means of evaluators. Result and discussion and be found later and conclusion section is at the end.

2   Challenges of Thai Tweet

Messages on Twitter are called 'tweets' and are restricted to 140 characters. Within this short length, users adopt a self-defined hashtags starting with '#' hash sign to give a specific defined word representing certain events or key purposes. The usage of hashtags are debated [1] to give rich meanings. Figure 1 gives an example tweet in Thai that comprises of common parts such as the general text, URL, HASHTAGS and MENTION (starting with '@' sign followed by a name of twitter user).

Fig. 1. A tweet structure

A concept of Twitter that helps spread out information quickly is the Retweet. When a tweet is retweeted, the original message is encapsulated within a new tweet (in the old version of Twitter, one can notice if the tweet are retweeted from the 'RT' character, however this is omitted in the current version). Normally users retweet because they agree with the message or want to promote the message. So, retweet can be regarded as an agreement or consensus on a particular information. It is known that processing Thai language is more difficult than languages in which word boundary markers are placed between words. Coupling with the limitation of 140 characters, Twitter users tend to create their own language in order to give more meaning. Thai slang and new terms are added via the usage of all social media, no exception for Twitter. This means it is even harder to process Thai language under such evolution of language. In this work, we will apply multi-applicable methods found working with other languages to extract key terms under such challenges of Thai tweet environment.

3 Related Works

Finding key features representing an event is known in literature as keyword extraction where originally working with search engine and document corpus [2]. Later, social media analytic has been adopting the similar algorithm and methods for the same purposes in the dynamic corpus and a near real-time environment. In our previous implementation, we use purely frequency of terms within a corpus (a set of tweet for an event) to suggest the good key terms that can represent event. However the literatures [3, 4] favor TFIDF as a feature weighting to emphasize the informative of such term over the corpus. Thus, we study TFIDF against term frequency (TF) for our objective. Actually, both of TF and TFIDF is feature vectorization method widely used in text mining to reflect the importance of a term to a document in the corpus. Denote a term by t, a document by d, and the corpus by D. Term frequency $TF(t, d)$ is the number of times that term t appears in document d, while document frequency $DF(t, D)$ is the number of documents that contains term t. If we only use term frequency to measure the importance, it is very easy to over-emphasize terms that appear very often but carry little information about the document, such as those articles found in stop words in English language. If a term appears very often across the corpus, it means it doesn't carry special information about a particular document. Inverse document frequency is a numerical measure of how much information a term provides: $IDF(t, D) = log(|D| + 1/DF(t, D) + 1)$, where $|D|$ is the total number of documents in the corpus. Since logarithm is used, if a term appears in all documents, its IDF value becomes 0. Note that a smoothing term is applied to avoid dividing by zero for terms outside the corpus. The TFIDF measure is simply the product of TF and IDF: $TFIDF(t, d, D) = TF(t, d) \times IDF(t, D)$.

In order to do Thai language processing, we need a Thai tokenizer. Haruechaiyasak and Kongthon [5] presents LexToPlus, a Thai lexeme tokenizer, that uses a longest matching dictionary-based approach with a rule-based normalization process. The work claims yield of accuracy at 96.3% on a test data set. To the best of our knowledge, we couldn't find other Thai tokenizer that yield better result than this one. So we utilize LexToPlus in this work.

4 Research Methodology

We construct 14 different approaches to select and weighting features of an event. The section will start with preprocessing data in details, followed by the explanation of feature selection and weighting schemes used in this work, then summarizes all 14 approaches in a figure.

4.1   Preprocessing Data

Since this study tries to comprehend the overall key terms of an event, we are attracted to the general text and the hashtag parts as shown in Fig. 1. Data preprocessing is needed to remove noisy and meaningless part out of tweet text. There are several preprocessing steps we used and can be described as follows.

Remove hyperlinks (URLs) Due to the fact that tweet has little space of 140 characters, most URLs in tweet are shorten URL which the real domain name are removed. Hence, the URLs in tweet cannot give a comprehensive information.

Remove special characters and emoticons Usually, punctuations and special characters are not meaningful for text mining. In some cases, like sentiment analysis, emoticons are useful, but not for the current study.

Remove retweet signs and twitter username A Retweet (RT) is a re-posting message of someone else's Tweet [6]. Former style of retweeting is indicated by letters 'RT' and twitter username (with an at sign '@' before the name) found at the beginning of message as shown in Fig. 2. Although, Twitter provides new official retweet feature that omit those explicit 'RT' and 'the original Twitter user', many users and many twitter client applications still apply the old style of retweeting. Since 'RT' doesn't give a comprehensive info to event, we eliminate both 'RT' and username that appear in a tweet.

Original Message: สรุปความเห็นส่วนตัว *#thevoiceTH* ซีซัน 4 เทปแรก สิงโต โค้ชคนใหม่สอบ ผ่าน โค้ชคนอื่นฮาเหมือนเดิม นักร้องได้มาตรฐาน มีสองสามคนที่โดดเด่น

Retweet Message: RT *@kidmakk:* สรุปความเห็นส่วนตัว *#thevoiceTH* ซีซัน 4 เทป<u>แรก</u> สิงโต โค้ชคนใหม่สอบผ่าน โค้ชคนอื่นฮาเหมือนเดิม นักร้องได้มาตรฐาน มีสองสามคนที่(...)

Fig. 2. Example of old style retweeting that cut the tail part of original tweet and replaced by ellipsis sign

Remove incomplete word The retweeted message by unofficial retweet feature will automatically add letters 'RT' and username of the original tweet message as explained earlier. The adder of such extra letters makes it possible to exceed the 140 characters limitation and thus the end of the original text might be replaced by an ellipsis sign (...). This often happens with the last word or the last hashtag of the message.

Normalization One common erroneous for text tokenizing in Thai social media text processing is the insertion word [5]. An insertion word is a word or term that have a repeated characters, e.g. veryyyyyy goooooooooooood or hellooooooooo. In this case, all insertion words will be normalized at this preprocessing step.

4.2   The Studied Approaches: Feature Selection and Weighting Schemes

This work focuses on studying 3 classes of feature type: **hashtag**, **unigram** and **bigram**. Hashtag is an interesting word form since it is what users defines themselves. Depending on the erroneous of the users' language, hashtag may or may not making sense to be used as representatives of the event or topic. This work would like to confirm the usefulness of hashtag. In our previous work, we also use hashtag. However, we suspect that longer word or longer hashtag term might get attention from reader and, as well, might be more informative than short term one. Thus, our hypothesis is that the popular hashtag terms may comprise of a number of characters higher than a threshold. Based on such hypothesis, we study the average character length of the corpus we used, LexToPlus, and the famous corpus for standard Thai language research, BEST [7]. BEST collects approximately 5 million thai terms from multiple formal document sources and it is the best known downloadable corpus for today. From our preliminary study on both corpuses, we found that LexToPlus gives an average term length at 5.28 characters, while BEST is at 10.79 characters. Since our context is about data on Twitter that normally use informal language, we decided to focus using 5 characters as the threshold of term length for our hashtag approach, henceforth is called as **Hashtag (5) (so called, hash-tag-five)**.

Following the same hypothesis about the higher length term may present a better meaning for audience, bigram might be able to convey the message clearer than uni-gram. Since Thai word can be considered as unsegmented language, a mixture between two consecutive unigrams will create a simple bigram and that solely depends on text tokenization and its dictionary. However the created bigram will convey a meaning or not depends on many factors. One issue we foreseen as a key factor is the existence of **stop words**. Stop words are common words frequently appear in texts. These words have low significant value in distinguishing topics or categories. Stop words include functional words such as prepositions, conjunctions and articles. In general, stop words are recommended to be discarded before executing the classification or language processing task [8]. Based on the goal of this work, we found some Thai words function as the definition of stop words shown above, but the POS of itself is not. These words can be categorized into head words as defined by [9]: the head of a phrase is the word that determines the syntactic type of that phrase. For example, the head of the noun phrase boiling hot water' is the noun water'. Example of head words that is categorized as stop words in this work can be found in the Fig. 3, such as 'broth-ers', 'sister' and 'news'. Henceforth, we called stop words in the way that already include these selected head words in Thai language.

Stop words and head words	Example Bigram with Meaning
รอบ (around)	รอบอนุสาวรีย์ชัย (around Victory Monument)
ร่วม (join, include, together, coordinate)	ร่วมกิจกรรม (join the event)
นำ (bring, lead)	นำขบวน (lead the parade)
ข่าว (news)	ข่าวอุบัติเหตุ, ข่าวระเบิด ({accident, blasting} news)
เป็น (is, am, are)	เป็นกำลังใจ (supporting)
ด้วยกัน (together)	คนไทยด้วยกัน (we are all Thais)
ใกล้ (near)	ใกล้ท่าเรือ, ระเบิดใกล้ (near dock, blast near)
พี่ (brother, sister)	พี่ก้อง, พี่คิ้ม, พี่โจ้, พี่สิงโต (brother Kong, sister Kim, brother Joe, brother Singto)
กัน (together)	รักกัน, รู้กัน, เข้าใจกัน ({love, know, understanding} each other)

Fig. 3. Example of stop words list and their possible bigram with meaning

So far, we address that stop words has no meaning and should be discarded. However, for Thai language, including stop words in some domain may generate bigram with meaning. Figure 3 shows sample set of stop words used in this work and their possible bigram that give example meaning. Thus, for bigram we study the affection of stop words towards the bigram model via two study approaches: bigram that includes stop words, named shortly as **wst.bigram** (with-stop-word bigram) and bigram without stop words, called **nst.bigram** (no-stop-word bigram).

As mentioned under Sect. 3, this work is interested in comparing feature weighting between TF and TFIDF. So our approaches include both TF and TFIDF into all study feature selection, which so far has Hashtag(5), **wst.bigram**, **nst.bigram** and unigram. Also, because our weighting methods focuses mainly on frequency of terms, it cannot help to think about the affection of retweeting over our selection method. The more retweeting of a particular message, the higher the TF value. In order to understand the effect of retweeting over the result ranked set, we distinct tweet used in this work per event as another feature selection on main selection methods like bigram and unigram. Thus, we have more selection method, including **nst.bigram.dist**, **wst.bigram.dist**, **unigram.dist**. Henceforth, we call Hashtag(5), unigram, wst.bigram, and nst.bigram as the main feature selections, with variance of distinction (.dist) and the feature weighting of TF and TFIDF. In total there are 14 studied approaches which are the combination of main feature selections with variance of distinction and feature weighting, summarized in Fig. 4.

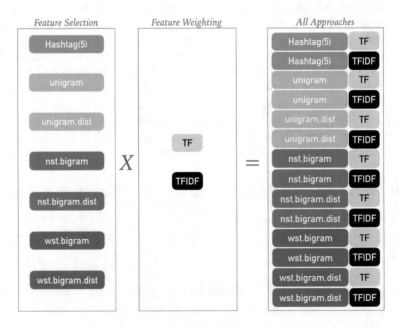

Fig. 4. All studied approaches

5 Experiment Settings

5.1 Data Set: Three Thai Tweet Events

The 3 different events used in this work occurred around AUG–SEP 2015. They comprise (1) the two consecutive bomb blasting in Bangkok at Ratchaprasong inter-section and Sathorn dock, named as **Ratchaprasong event**; (2) the **Bikeformom event** is a cycling event to mark Her Majesty the Queen's 83rd birthday; (3) the Voice Thailand is a singing contest television show aired around the beginning of September every year, we called it as **Thevoiceth event**. All events has different characteristics such as the number of days the audience interested and participated in the event, the number of tweet over the interest period, the type of event, etc. We can elaborate more about each event as follow.

For Ratchaprasong event, it is an emergency, fierce and can be regarded as a small crisis for Bangkok people. After the first bomb at Ratchaprasong junction (17th August 2015), the second bomb occurred at the crowded Sathorn dock, one of the main dock for commuting over the Chaopraya river (18th August 2015). These consequent bombing events make people feel panic and insecure mainly for people inside Bangkok area. Nevertheless, the event matches very well with the behavior of Twitter users that tend to spread the crisis news quickly and keep up-to-date for every information coming to their hands. Apart from giving and gathering information, twitter users also share their feelings and some tried to find hints about who is the bomber.

On the other hand, Thevoiceth is an entertaining event that happens once a week on Sunday evening during September-to-December each year. This year is the forth of the series which can guarantee the popularity of this TV program in Thailand. Twitter users give interest to this event since it started 4 years ago, and these users tend to tweet in order to support the singer or the coach they prefer. Normally, the related-Thevoiceth tweet will increase sharply from Saturday to Sunday evening in which the show is on-air.

The Last event, Bikeformom is a special event that never happens before and can be regarded as a general event to twitter user that happens in society only once. Although, it is a big event to mark Her Majesty the queen birthday, but for twitter users, there is no much information to spread out or any particular issue or trend to be reported. This is due to the main channel of publication has already been placed by free TVs, Newspapers and other channels. So, mainly twitter users tweet about their mutual feeling toward the Bikeformom event in different point-of-views and other small related issues. The Bikeformom event comprises two major related moments that are the Thai Mother day (August 12, 2015 or Her Majesty the Queen birthday) and the cycling day (August 16, 2015).

5.2 Implementation

The implementation in this work has two parts. One is a web application for surveying vote from evaluators. The second part is the engine for our studied approaches. Here we will discuss the implementation about the engine part of this work.

Since text processing is a high computational task. Coupling with the amount of tweet data to process is huge and dynamic in general. We are attracted to Apache Spark as our solution for scalable and time-saved operation. Apache Spark [10] is one of the popular general engine that introduces the concept of resilient distributed datasets (RDDs) to enable fast processing of large volume of data leveraging distributed memory. In-memory data operations makes it well-suited for iterative applications such as iterative machine learning and graph algorithms. Spark also provides MLlib that is Apache Spark's scalable machine learning library. It contains high-quality algorithms that leverage iteration, and can yield better results than the one-pass approximations sometimes used on MapReduce.

Our implementation preprocesses data and utilizes a ready-to-use TFIDF from MLlib. The implementation is done with Java and Scala using MapReduce concept on Apache Spark platform. Moving implementation to MapReduce concept has a challenge in itself, since the coding style need to be changed in order to gain speed and other facilities of distributed computing. All study approaches will be feed with each event set of tweets and return a set of terms to be represented for such event. In this work, we implement the experiment with the intention to reuse the best approach in the real world application. Thus, the implementation is ready to be used in PaaS or Spark Cluster Platform which works in distributed manner. The speed of this implementation with Spark is better than the non-distributed version. For instance, the *nst.bigram.dist. tfidf* approach, which is the most complex among others, is used with the data of 3.5 million tweets of Ratchaprasong event on a single machine (MacBook Pro 2012 model) with RAM 8G and it gave the result within 10 minutes. Actually, within such 10 minutes, it includes all preprocessing data stated in Sect. 4.1, feature selection, weighting scheme and selecting 20 best key terms for the event. By using the same preprocessing step plus the algorithm, our previous work with a non-distributed implementation gave us result within two hours.

5.3 Validation Methodology

In order to validate the quality of terms given by each studied approach, we need an evaluation from human to compare with. Thus, the top 50 terms resulting from each approach are added into an evaluating preference survey by volunteers. With the survey web application, the evaluator starts by giving his/her basic demographics, choosing the event to be evaluated and, in a per event manner, selecting up to 20 most relevant terms to explain the event. The evaluators are asked to select only terms, they think, that are the best to capture the event. Also there is no need to fill in non-relevant words to reach 20 terms. After weeks of surveying period, there are 162 evaluators in total in which 89% are between 20–46 years old and 74.6% use social media for 3 hours or more per day. The survey gives the vote number per terms from our evaluators. Then, the vote frequency of each term are gathered, ranked, plotted and analysed.

It is important to note that when we include 50 terms from all approaches, we found that many approaches suggest the same term. We decided to use distinct terms to be included in the survey, so that evaluators will not get confuse by repeated terms. Thus, it means each of these selected terms might be tagged with more than one approach that suggests it.

Since our goal is the find the best approach that can give the best 20 set of terms capturing the idea of event on twitter data. The performance can only be confirmed by comparing to what evaluators think what terms to represent the event. However, each of these survey terms may be selected by multiple approaches. Thus, it is possible to have many approaches that can select the best set of terms. We will further discuss these 20 top voted terms along with their generated approaches event-by-event in the next section. As well as, we will compare and discuss the best selected terms given by the best approaches with the original baseline in the next section too.

6   Result Discussions and Perspectives

First glance at the top 20 voted terms from all approaches and all event surprise us because there are so little differences between TF and TFIDF as well as the distinction and no distinction of retweeted message. For example, a term of 'Coach Kong' is given by (1) nst.bigram.tf, (2) nst.bigram.tfidf, (3) nst.bigram.dist.tf and (4) nst.bigram.dist. tfidf. To confirm that there is no differences among them, we do statistical analysis using Wilcoxon signed rank test [11] with continuity correction and found that the features selected by TF and TFIDF have no significant differences. Likewise, the tweet distinction feature (.dist) can be confirmed of no differences by the same method. By the confirmation of the Wilcoxon signed rank test above, we consider to ignore those TF/TFIDF (.tf and .tfidf) and tweet distinction features (.dist) and will analyse the results further based on the main feature selection approaches of (1) **Hashtag(5)** (2) **wst.bigram** (3) **nst.bigram** and (4) **unigram**.

Figures 5, 6 and 7 illustrate the top 20 voted terms from three different events (Ratchaprasong, Thevoiceth and Bikeformom, accordingly) by 162 evaluators in the stack bar chart. For the set of circles residing beside the bar chart, it shows the number of terms suggested by different experiment approaches. The set of circles also illustrates the number of overlapping terms by different approaches. These overlapping terms are interesting to be used as a measurement if one approach can be used instead of the others. When we give an overview on the top of all three Figs. 5, 6 and 7, we found that the most voted terms (the highest three terms) come from Hashtag(5) approach with very high frequency of evaluators when compare to the other approaches. While unigram approach has been chosen with the least frequency.

Considering the set of circle on Fig. 5, Ratchaprasong event grasp our attention since it has the biggest set of Hashtag(5) terms appearing in the result for 13 out of 20. While nst.bigram, wst.bigram and unigram are given 5, 3 and 3 terms respectively. Similarly for Thevoiceth event, Hashtag(5) term is also selected the most at 10 terms and all of them are at the top of the list, following by 9 altogether bigram and 2 terms of unigram. However, it is very obvious that Bikeformom event differs from the rest. It has the big group of bigram at 9 terms, followed by unigram and Hashtag(5) at 7 and 5 terms respectively. We discuss the reason of this finding later.

It is interesting to note that both wst.bigram and nst.bigram give very similar set of results in each event, as readers can see from the overlapping area on each figure. However, their majority of being superset or subset are vary among events. So this is the main reason that we cannot conclude what is the best approach between wst.bigram

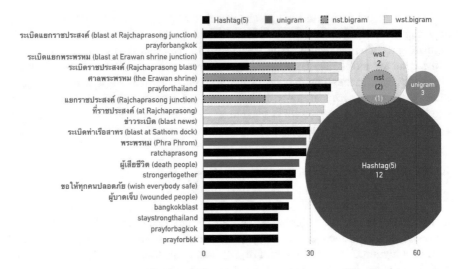

Fig. 5. The top 20 voted terms representing the Blast events in Bangkok, August 2015, and the number of term proportions selected by each extractor

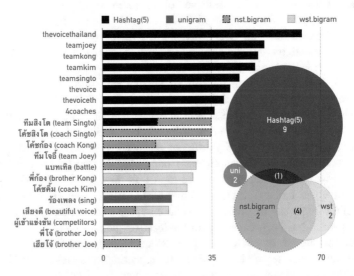

Fig. 6. The top 20 voted terms representing the Voice Thailand 2015 and the number of term proportions selected by each extractor

or nst.bigram to be used for suggesting the best key terms. From the result, it is undeniable, that among our approaches, Hashtag(5) is the best candidate approach according to the frequency of vote. Though it is not so obvious for the event like bikeformom in terms of voted terms proportion. Hence we observed the behaviors of hashtag over all experiment events using some statistics. The results are shown in Table 1.

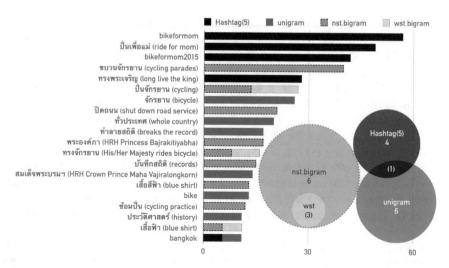

Fig. 7. The top 20 voted terms representing the Bike for mom 2015 and the number of term proportions selected by each extractor

Table 1. Statistics of tweet and hashtag per event

Events	Number of tweet	Number of distinct hashtag	Ratio of distinct hashtag over all tweets (DHOT)	Number of all hashtag used	Reusing hashtag ratio (RHR)
Bikeformom	272,185	3582	1.32	346,073	96.45
Thevoiceth	246,627	2362	0.96	494,781	209.48
Ratchaprasong	3,551,593	12,780	0.36	3,544,160	277.32

In general, Twitter users try to differentiate their tweet through hashtag manipulation. This usually makes hashtag carrying information that relevant to a particular event or a specific purpose. As a result, Twitter readers can recognize well such informative hashtags and tend to reuse such hashtags since it can communicate their thought. With the retweet concept, the well captive hashtag will be disseminated widely further.

From the Table 1 We found that the ratio of Distinct Hashtags Over all Tweets (DHOT) and Reusing Hashtag Ratio (RHR) can indicate some interesting characteristics. For DHOT, it shows how specific or broad meaning of hashtags given in particular event. The less the value of DHOT, the more specific or focus of hashtag terms used towards a particular meaning. For RHR, if the value is high, the hashtag is somehow reusable or convey the important message for the event. Hence, it has been reused again and again. It can be said that if hashtag can do its job very well, DHOT should has low value and RHR value should be high.

The Bikeformom event is interesting regarding this issue. For example, Bikeformom has the higher number of tweet and DHOT than Thevoiceth event, but the value of RHR of Bikeformom is less than Thevoiceth (95.45 and 209.48 accordingly). This comply to the vote given by evaluators where the hashtag is selected more in the case of Thevoiceth event than found in Bikeformom, as shown in Figs. 6 and 7. In other words, hashtags of Thevoiceth can convey more message about the event than what hashtags of Bikeformom do. Moreover, when we compare the value of DHOT and RHR together with the number of selected hashtag from all events; they all conform.

Thus, an interesting finding of this work is about the affection of the event characteristics towards how to select the representative key terms per event. Ratchaprasong can be categorized into an emergency and crisis event. As shown in Table 1, the RHR value of Ratchaprasong event is the highest among the rest, 13 hashtags are selected by evaluators. The second places falls for bigram, both with and without stop words, in which bigram terms are mostly related to name of places. So, we notice that place-related-term is important for this particular emergency and crisis event type. On the other hand, Thevoiceth is an entertainment periodic event (aired on every Sunday evening for 4 consecutive months a year). Tweets for Thevoiceth rise highly around the on-air time. Normally this kind of event such as weekly based drama (e.g. Game of Throne) have similar pattern of tweets like Thevoiceth. From the result, we notice that top voted terms give basic information about the event, i.e. coach's name, rules and activities of the competition. These information does not change much from week to week. The last event is Bikeformom. This event is a special event at nation wide level. The event takes time around two months starting from the publicity of the event to the cycling date. However, highly tweets occur only around two days of Her Majesty the Queen's birthday and of the cycling day. Also, we notice that the event hold very specific character so the hashtag usage repeatedly is very narrow to the name of the event ('bikeformom', 'ride for mom' and 'bikeformom2015'), and the phrase 'long live the king'.

Another issue is about tokenization we used, LexToPlus. Since it is the longest matching tokenizer with dictionary based, some unigram terms can be categorized as bigram word form. For example, the term 'breaking the record' presented in Thai can be split into 'destroy' and 'statistic'. This occur because inside dictionary has 'destroy', 'statistic' and 'breaking the record' and the longest matching tokenizer will choose the longest token to be used first. For this finding, we foreseen that generalized definition of unigram and bigram inside Thai NLP society is needed. However, in this work we rely on the result from LexToPlus and it evidently turns out that unigram approach has been selected the least among the others as shown in all result figures.

Lastly, in order to confirm the performance of the resulting approaches, we provide an example of terms given by the previous work as baseline comparing with our proposed methods. Figure 8 illustrates examples of the hashtag-baseline and selected terms by Hashtag(5), our method, to present on our application. While, Fig. 9 shows the terms from keyword-baseline, together with, wst.bigram and nst.bigram. Both figures used tweet from Ratchaprasong event in which we conclude to use both Hashtag(5) and all bigram series.

hashtag-baseline	Hashtag(5) approach
ระเบิดราชประสงค์ (bombing at Rajchaprasong)	prayforbangkok
prayforbangkok	strongertogether
strongertogether	ระเบิดราชประสงค์ (bombing at Rajchaprasong)
ประเทศกูรักสะอาด (my country loves clean)	prayforthailand
bangkokblast	bangkokblast
ราชประสงค์ (Rajchaprasong)	prayforbagkok
prayforthailand	ราชประสงค์ (Rajchaprasong)
prayforbagkok	bangkok
มีความหวังละ (got hope)	เราจะเติบโตและแข็งแกร่งไปด้วยกัน (we'll grow stronger together)
bombercluephoto	bombercluephoto
ระเบิด (bomb)	staystrongthailand
prayforenglish	ระเบิดแยกพระพรหม (bombing at Brahma shrine)
staysafebangkok	ระเบิดแยกราชประสงค์ (bombing at Rajchaprasong junction)
thaipbsnews	ratchaprasong
rip	staysafebangkok
tianjin	thailand
stronger	welovethailand
r.i.p.	ระเบิดท่าเรือสาทร (bombing at Sathorn dock)
หืม (hmm)	prayforbkk
คนไทยและชาวต่างชาติมาร่วมจุดเทียน (both Thais and foreigners light candles together)	erawanshrine

Fig. 8. Previous key hashtags compares with the proposed approaches

keyword-baseline	nst.bigram approach	wst.bigram approach
ระเบิด (bomb)	แยกราชประสงค์ (Rajchaprasong junction)	เหตุระเบิด (bombing)
ราชประสงค์ (Rajchaprasong)	ระเบิดราชประสงค์ (bombing at Rajchaprasong)	ระเบิดที่ (explode at)
พระพรหม (Brahma)	ท่าเรือสาทร (Sathorn dock)	แยกราชประสงค์ (Rajchaprasong junction)
สาทร (Sathorn)	เกิดเหตุระเบิด (bombing happens)	เกิดเหตุระเบิด (bombing happens)
เหตุการณ์ (event)	ศาลพระพรหม (Brahma shrine)	ศาลพระพรหม (Brahma shrine)
มือระเบิด (bomber)	แสดงความเสียใจ (express condolence)	ท่าเรือสาทร (Sathorn dock)
ตำรวจ (police)	แรงระเบิด (power of explosion)	อยากให้ (wish)
ผู้ต้องสงสัย (suspected person)	ปาระเบิด (throw bomb)	แสดงความเสียใจ (express condolence)
ท่าเรือ (dock, pier)	ระเบิดแสวงเครื่อง (improvised explosive device)	ระเบิดแยก (explode at junction)
ทหาร (soldier)	บริจาคเลือด (blood donation)	เกิดระเบิด (bombing happens)
คนร้าย (villian)	มือวางระเบิด (bomber)	คนวางระเบิด (bomber)
สะพาน (bridge)	เก็บกู้ระเบิด (dispose bomb)	ข่าวระเบิด (bombing news)
เกาหลีใต้ (South Korea)	สะเก็ดระเบิด (shrapnel)	เจอระเบิด (found bomb)
เกาหลีเหนือ (North Korea)	วัตถุต้องสงสัย (suspected object)	มีระเบิด (have bomb)
เกิดเหตุ (happen)	จุดระเบิด (ignition)	แรงระเบิด (power of explosion)
เจ้าหน้าที่ (officer)	ปิดถนน (close road)	ไปด้วยกัน (go together)
บ้าน (house)	วางพวงมาลัย (place garland)	ที่ราชประสงค์ (at Rajchaprasong)
ทุกคน (everyone)	แชร์รูป (sharing image)	ปาระเบิด (throw bomb)
ผู้เสียชีวิต (the deceased)	จุดเกิดเหตุ (crime scene)	เหตุการณ์ระเบิด (bombing event)
ต่างชาติ (foreigner)	ภาพสเก็ต (sketch image)	มีคนเจ็บ (has casualty)

Fig. 9. Previous key features compares with the proposed approaches

Furthermore, we do F-score analysis to confirm the performance of the selected terms by our approaches with what evaluators choose. This comparison can be seen in Table 2 and it is obvious that our approaches give a better performance as F-Score value is all higher twice as much.

Table 2. The F-Score comparison between the baseline and our approaches on Ratchaprasong tweet

Approaches	Precision	Recall	F-score (F1)
Hashtag-baseline	0.14	0.12	0.13
Hashtag(5)	0.25	0.20	0.23
Keyword-baseline	0.06	0.06	0.06
nst.bigram	0.12	0.11	0.12
wst.bigram	0.12	0.12	0.12

7   Conclusion

The target application of this work wants to show a good-enough set of key terms representing any event occurring on Thai Twitter society. So, this work finds appropriate approaches to realize such need. Our studied approaches covered preprocessing, feature selections and weighting schemes on three Thai real tweet events. We evaluate the performance of approaches by means of survey with 162 evaluators. To conclude, firstly, we found that the characteristics of event plays an important role on how to consider the proper approaches. After considering the performance among Hashtag(5), nst.bigram, wst.bigram and unigram on all events, our conclusion is to use the combination of all preprocessing together with Hashtag(5) and bigram to provide the most suited hashtags and keywords for the event similar to Thevoiceth and Ratchaprasong. On the other hands, we tend to use nst.bigram, wst.bigram and unigram for the event like Bikeformom. Second, we confirm the usefulness of stopwords in Thai language, as we find that wst.bigram can contribute key terms for each event and is preferable by evaluators.

Third, we found no statistical differences between weighting schemes of TF and TFIDF, as well as, the distinction of tweet (.dist). Thus, we suggest TF since it uses lower computational time without loosing quality of the result terms. Forth, We also found that the utilization of the average character length found in Lextoplus as the threshold for our Hashtag(5) approach is working very well. From evaluators' vote, the average character length is 12.5, 7.8, 9.1 for hashtag, wst.bigram and nst.bigram, respectively. This information can confirm our hypothesis about the long character term can grab more intention to tweet and give more informative than those short term: lower than 5.28 characters.

Our next focuses fall into categorizing 'tweet event type' and extract its key characteristics as well as audience's needs for each event type. So that the suggested key terms will fit users' need more. For example, the audience of Thevoiceth might want to know who sing what song for the week and how well they do. The expectation has shifted from the common info of the event to the real content of that particular week. Moreover, the near real-time processing tweet event should be explored. For the Thai NLP, differentiating stop words and/or head words of each event type may help increasing key terms selection process.

References

1. Hong, L., Davison B.D.: Empirical study of topic modeling in twitter. In: Proceedings of the First Workshop on Social Media Analytics, pp. 80–88 (2010)
2. Kaur, J., Gupta, Vishal: Effective approaches for extraction of keywords. J. Comput. Sci. **7** (6), 144–148 (2010)
3. Abilhoa, W.D., De Castro, L.N.: A keyword extraction method from twitter messages represented as graphs. Appl. Math. Comput. **240**, 308–325 (2014)
4. O'Connor, B., Krieger, M., Ahn, D.: TweetMotif: exploratory search and topic summarization for twitter. In: 4th International AAAI Conference on Weblogs and Social Media, pp. 2–3 (2010)
5. Haruechaiyasak, C., Kongthon, A.: LexToPlus: a Thai lexeme tokenization and normalization tool. In: The 4th Workshop on South and Southeast Asian NLP (WSSANLP), International Joint Conference on Natural Language Processing, pp. 9–16. Nagoya, Japan (2013)
6. Inc. 2015 Twitter. https://support.twitter.com/articles/77606
7. LST, NECTEC. BEST: http://thailang.nectec.or.th/downloadcenter/ (2016)
8. Sukhum, K., Nitsuwat, S.: Opinion detection in Thai political news columns based on subjectivity analysis. In: The 7th International Conference on Computing and Information Technology IC2IT2011, pp. 27–31 (2011)
9. Zwicky, A.M.: Heads, bases, and functors. In: Heads in Grammatical Theory, pp. 292–315. Cambridge University Press (1993)
10. Apache Spark. http://spark.apache.org/ (2015)
11. R-tutor. http://www.r-tutor.com/elementary-statistics/non-parametric-methods/wilcoxon-signed-rank-test (2016)

Comparison of Document Clustering Methods Based on Bees Algorithm and Firefly Algorithm Using Thai Documents

Pokpong Songmuang[✉] and Vorapon Luantangsrisuk

Department of Computer Science, Thammasat University, Bangkok, Thailand
pokpong@cs.tu.ac.th, voraponl@sci.tu.ac.th

Abstract. Several researches performed experiments to compare performances of data clustering methods based on nature-inspired algorithms using various data types including text document dataset. According to the results, although Firefly algorithm showed high performance to cluster text document over several nature-inspired algorithms but Bees algorithm had better performance to solve complex problems than several nature-inspired algorithms. However, none of the experiments compared the performances of the both algorithms to cluster text documents. Therefore, we compare the document clustering performances of the clustering methods based on Firefly algorithm and Bees algorithm.

Keywords: Document clustering · Firefly algorithm · Bees algorithm · k-means

1 Introduction

Text document clustering plays an important role in modern data analysis since the online information is increasing exponentially day to day. Text document clustering is necessary process, such as to organize massive amount of documents into meaningful clusters. Each cluster is a collection of documents that are similar to on another within the same cluster and are dissimilar to documents in other clusters. This clustering is important in many data mining research areas, including text mining, information retrieval, web search, and etc.

k-means clustering [1] is one of the most popular methods for text clustering since k-means clustering is very simple to implement and has a computational efficiency. However, the main problem of k-means clustering is that it easily converge to local optima [2].

To solve this problem, several research apply nature-inspired algorithms to cluster text documents, for example, Genetic Algorithm (GA) [3, 4], Particle Swarm Optimization (PSO) [5, 6], Ant Colony Algorithm [7], Bees Algorithm (BA) [8], Firefly algorithm (FA) [9–11].

Mohammed et al. [10] proposed the document clustering method based on FA with k-means and performed some experiments to show that the performance of the proposed method was better than the document clustering methods based on the basic k-means and the other nature-inspire algorithms.

T. Theeramunkong et al. (eds.), *Advances in Natural Language Processing,*
Intelligent Informatics and Smart Technology, Advances in Intelligent Systems

AbdelHamid et al. [8] and Mizooji et al. [12] proposed data clustering methods based on BA including document clustering method. They also performed experiments to compare the performances of the proposed methods with the other nature-inspired algorithm and also k-means. According to the results, the performances of the methods [8, 12] based on BA were better than the other methods and also k-means. However, FA is not included in the both researches.

In this paper, we are interested in the performances of document clustering methods based on nature-inspired algorithms. Therefore, we perform some experiments to reveal the document clustering performances between BA and FA in which the experiment has never done yet.

Next section, we explain the related works about methods for text document clustering. We also explain how to apply BA and FA to cluster text document data. Then we describe the details of experiment to compare the performances of the text document clustering methods and results of the experiments.

2   Related Works

2.1   Document Representation

Most clustering algorithms represent the dataset as a set of vectors, where the vector corresponds to a single element and is called a feature vector. The feature vector contains appropriate features to represent the element. For text document, elements are documents and terms in documents represent the features of documents. Therefore, the text documents can be represented with the vector of terms which call Vector Space Model (VSM) [13]. In this model, a document is represented by a vector d, such as $d = w_1, w_2, \ldots, w_m$, where $w_i(i = 1, 2, \ldots, m)$ is the term weight of the term t_i in one document. The term weight value represents the significance of this term in a document. To calculate the term weight, the occurrence frequency of the term within a document and in the entire set of documents must be considered. The most widely used weighting scheme is the Term Frequency - Inverse Document Frequency (TF-IDF) [14]. The weight of term i in document j is given in Eq. 1:

$$w_{ji} = tf_{ji} * idf_i \tag{1}$$

where tf_{ji} is the number of occurrences of term i in the document j; idf_i represents how common or rare is term i across all documents. idf_i can be calculated by the following equation:

$$idf_i = \log_2 \frac{N}{df_i} \tag{2}$$

where df_i is the number of documents in the collection which term i occurs; and N is the total number of documents in the collection.

In this paper, we represent each document as the term weight vector. In the next section, we explain k-means algorithm for text document clustering.

2.2   *k*-Means

k-means algorithm [1] is one of the most popular data clustering algorithm since it is simple to implement and has a computational efficiency.

The *k*-means algorithm clusters a group of data into a predefined number of clusters, K. The *k*-means algorithm can be summarized as:

1. Randomly select K documents as a set of cluster centroids to set an initial dataset partition, $c_1, c_2, \ldots, c_K$. Here, the document d and the centroid c_k are represented by the term weight vectors described in Sect. 2.1.
2. Assign each document, d, to the closest cluster centroid, c_k, using Euclidean distances, D, as the following equation:

$$D(d, c_k) = \sqrt{\sum_{i=1}^{m} (w_i^d - w_i^{c_k})^2}. \qquad (3)$$

3. Recalculate the cluster centroid c_k based on the documents assigned to the centroid, c_k, using the equation bellow:

$$c_k = \frac{1}{n_k} \sum_{\forall d \in S_k} d, \qquad (4)$$

where c_k is the centroid of cluster S_k; d denotes the document that belongs to cluster S_k; c_k represents the centroid; n_k is the number of documents that belong to cluster S_k.

4. Repeat step 2 and 3 until the stopping criterion is met.

The stopping criterion is the maximum number of iterations or when the result of cluster centroid recalculation no longer change.

Although the *k*-means algorithm is simple and popular, the main drawback is that it easily converge to local optima. Therefore, several research apply the nature-inspired algorithms to relax the problem. We describe the text clustering methods based on BA and FA in the next sections.

2.3   Bees Algorithm

Bees in Nature Honey bees live in a hive where they store honey that they have foraged. Honey bees can communicate the locations of food sources to their hive mates by performing a so-called "waggle dance". The durations of this dance are proportional to the quantities of food at the sources. By engaging in this behavior, large groups of bees are recruited to forage sources that contain large quantities of food. This reduces the individual time required to forage for food [15].

Bees Algorithm for Optimization Problems BA is an optimization algorithm inspired by the natural foraging behavior of honey bees to find the optimal solution [16]. BA steps are given below:

1. Initial population of artificial bees randomly search for solutions.
2. Evaluate fitness of the solutions.
3. Continue below steps until stopping criterion is met
    (i)   Choose the solutions with highest fitnesses and allow them for neighborhood search.
    (ii) Recruit more artificial bees to search near to the high fitness solutions and evaluate the fitnesses.
    (iii) Select the best solution from each patch.
    (iv) Remaining bees are assigned for random search and do the evaluation.

In order to cluster data using bees algorithm, we must first model the clustering problem as an optimization problem that locates the optimal centroids of the clusters rather than to find an optimal partition. The following subsections describe the proposed Bees Algorithm for text document clustering.

Bees Algorithm for Text Document Clustering In order to apply BA to cluster text documents, we develop the text document clustering method based on the data clustering method based on BA [12]. From here, the text document clustering based on BA is called TDCBA.

At first, we define the number of clusters, K. Here, the documents are represented using the vector of weight of terms and use floating-point vectors to represent cluster centroids, c_k. Each document must be assigned exactly to one cluster, S_k. For each artificial bee has an assignment that is a set of cluster centroids with K nonempty clusters. In this model, each food source discovered by each artificial bee is a candidate solution and corresponds to a set of K centroids.

Next, all artificial bees randomly select K documents as the centroids of clusters and assign each document, d, to the closest cluster centroids.

Then we evaluate the fitness values, fit, of the solutions from artificial bees using the internal cluster distance as shown in the equation below:

$$fit = \frac{1}{\sum_{k=1}^{K} \sum_{\forall d_k \in S_k} D(d_k, c_k)} \tag{5}$$

where D is Euclidean distance as shown in Eq. 3 and our objective function is to minimize the internal cluster distance.

The assignments are ranked by fitness values and they are assigned the selection probabilities calculated from the fitness values. The selection probability of assignment is directly proportional to the fitness value of each solution.

Next, the assignment are randomly selected based on the selection probabilities and are assigned to the artificial bees for neighbourhood search. After the artificial bees found the candidate solutions, they comeback to store the solutions and we start the next iteration. These steps are repeated until a stopping criterion is met.

In this problem, the goal is to find the maximum of the objective function.

2.4   Firefly Algorithm

Firefly in Nature In nature, male fireflies use flashing to attract female fireflies. The pattern of flashes are difference among firefly species. The flashing light of firefly is produced by a process of bioluminescence, and several issues such signaling systems are still debating [15].

Important common characteristic of flashing light of all fireflies is that the intensity of flashing inversely increases with distance between fireflies since air absorbs light. So a closer firefly is looked brighter than the farther firefly with the same flash light intensity. This idea is applied to develop Firefly Algorithm.

Firefly Algorithm for Optimization Problems Firefly Algorithm (FA) is a metaheuristic algorithm, inspired by the flashing behavior of fireflies [17]. The primary purpose for a firefly's flash is to act as a signal system to attract other fireflies. Yang [17] formulated this firefly algorithm by assuming:

1. All fireflies are unisex, so that one firefly will be attracted to all other fireflies.
2. Attractiveness is proportional to their brightness, and for any two fireflies, the less brighter one will attract (and thus move) to the brighter one; however, the brightness can decrease as their distance increases. The attractiveness function [17] is as follows.

$$\beta = \beta_0 * e^{-\gamma r_{ij}^2} \tag{6}$$

where β_0 is attractiveness at distance $r = 0$; γ is the absorption; $\gamma \in [0, \infty)$; and r_{ij} is distance between firefly i and firefly j.

3. If there are no fireflies brighter than a given firefly, it will move randomly.
4. The brightness should be associated with the objective function.

Recent studies show that FA is particularly suitable for nonlinear multimodal problems.

Firefly Algorithm for Text Document Clustering Here, we describe the text document clustering based on FA (TDCFA).

First, we define the number of clusters, K. Here, the documents are represented using the vector of weight of terms as described in Sect. 2.1 and we use floating-point vectors to represent cluster centroids, c_k. Each document must be assigned exactly to one cluster, S_k. Each firefly is a candidate solution which represents a set of cluster centroids with K nonempty clusters.

Next, all fireflies randomly select K documents as the centroids of clusters and assign each document, d_k, to the closest cluster centroids using Euclidean distance in Eq. 3.

Then we calculate the brightness of each firefly. Here, the brightness of firefly is represented by the fitness value of the candidate solution as shown in Eq. 5.

After that each firefly calculates the attractiveness of all fireflies based on Eq. 6 and ranking the attractiveness of fireflies. Here, the attractiveness is proportional to the brightness of firefly and absorption from the distance. The distance between fireflies can be calculated from the difference of the fitness values between two fireflies.

If the firefly meets the brighter firefly, then the firefly moves toward to the brighter firefly and find the new solution based on the solution of the brighter firefly, else the firefly moves randomly. These steps are repeated until a stopping criterion is met.

For example, Mohammed et al. [10] proposed the TDCFA with k-means. Mohammed et al. [10] also performed some experiments to show that the document clustering performance of the proposed method was better than the document clustering methods based on the basic k-means and the other nature-inspire algorithms.

Although FA and BA are famous and perform high performance to cluster data, none of previous researches performs an experiment to compare efficiency of the method based on BA and FA to cluster document. Therefore, we perform the experiments to compare the document clustering efficiency between them.

3 Experiment and Results

In this experiment, we compare the performances of TDCBA and TDCFA using data from Thai Thesis Database [18]. For more details, we compare the performances of three clustering methods as follows: (1) k-means, (2) TDCBA described in Sect. 2.3 and (3) TDCFA described in Sect. 2.4. All the text document clustering methods are implemented in java.

The details of data are as follows: (1) Thai abstracts from three domains including 3000 documents, (2) 1000 documents per domain, (3) 156541 sentences and (4) 22002 unique words. In order to reduce the bias results since all text document clustering methods in this experiment is based on random value, we perform experiment 10 times. Each time we randomly select 200 documents from each domain as an experiment data.

Since Thai sentences have non-breaking space which are different from English sentences, we have to prepare the data properly before we perform the experiment. Therefore, we perform word segmentation in which sentences are separated in words using a software called LexTo [19]. After that we convert the documents to vector space using TF-IDF as describe in Sect. 2.1 and perform text document clustering methods described in Sects. 2.2–2.4.

In this experiment, the maximum iteration numbers are defined as stopping criteria. The maximum iteration numbers and the numbers of population of artificial bees and fireflies are shown in Table 1.

Table 1. Number of population and maximum number of iterations

Parameter	TDCBA	TDCFA
Population	20	20
Max iteration	100	100

To evaluate the performance of text document clustering, we use F-Measure. The F-Measure index is used to numerically evaluate the efficiency of text document clustering methods. This index utilizes two concepts of precision and recall which are calculated by the following equations:

$$precision(p,q) = \frac{x_{pq}}{x_q} \tag{7}$$

and

$$recall(p,q) = \frac{x_{pq}}{x_p} \tag{8}$$

where x_{pq} is the number of data in domain p that in cluster q, x_q is the number of data in cluster q, x_p is the number of data in domain p.

Hence, the F index for class p and cluster q is calculated as follows:

$$F(p,q) = 2\frac{recall(p,q) * precision(p,q)}{recall(p,q) + precision(p,q)} \tag{9}$$

This is a value between 0 and 1, which a value near 1 is desirable. The average of precision and the average of recall obtained from k-means, TDCBA and TDCFA. They are presented in Table 2.

Table 2. The precision and recall from three methods using three domains of documents

Parameter	Precision			Recall		
	k-means	TDCFA	TDCBA	k-means	TDCFA	TDCBA
Domain A	0.3840	0.4514	0.9327	0.4090	0.4440	0.9570
Domain B	0.3812	0.3948	0.9580	0.3130	0.2760	0.7516
Domain C	0.4789	0.4573	0.8403	0.5335	0.6025	1

The underlines indicate the best values which imply the best performance method for each Domain

According to the precision results in Table 2, the text document clustering method based on BA performs the best precision. This means that this method clusters only few data into the group of majority documents form the diffidence domain. TDCFA and k-means perform similar precision performance.

On the other hand, according to the recall results in Table 2, TDCBA performs the best recall. This means that the method correctly clusters the documents from the same domain into the same group. Moreover the recall equals to one reveal that TDCBA rubust to cluster this kind of data. TDCFA and k-means perform similar recall performance.

Moreover we calculate the averages of the F-Measure indexes from the three methods using three domains of documents. The averages are show in Table 3.

Table 3. The averages of F-measure indexes from three methods using three domains of documents

Parameter	k-means	TDCFA	TDCBA
Domain A	0.3961	0.4477	0.9447
Domain B	0.3438	0.3249	0.8424
Domain C	0.5047	0.5200	0.9132

The underlines indicate the best values
which imply the best performance method
for each Domain

The results in Table 3 show that, for all domain of documents, TDCBA performs better performance than the other methods.

The results of k-means in Table 3 is the worst and TDCFA performance is slightly better than k-means. It seems k-means and TDCFA suffer from local optima. Although Hassanzadeh and Meybodi [9] show FA has better performance to cluster data than other algorithms, the experiment data has smaller number of dimensions than our experiment data. For more details, it is a large number of dimensions so distances between fireflies become large. Then they move randomly more than move toward the firefly with the better solution. Therefor the cluster centroids are moved randomly and TDCFA cannot reach the global optimum.

In more details, the results of Domain B of all methods are the worse because the data form Domain B contains more unique words than the other domains and has low frequencies for each word. In the other words, Domain B has more sparse data than the other domains so the results of precision, recall and F-Measure of clustering methods become the worse.

4 Conclusion

In this paper, we compared the performances of the text document clustering methods as follows: k-means, text document clustering method based on BA and FA. We compared the performances of the methods using three domains of Thai documents. According to the result of the experiment, the text document clustering method based on BA is better than k-means and the text document clustering method based on FA.

For future works, we plan to increase the number of domains and the number of data in order to compare the clustering performance of all methods in more details.

References

1. Macqueen, J.: Some methods for classification and analysis of multivariate observations. In: 5-th Berkeley Symposium on Mathematical Statistics and Probability, pp. 281–297 (1967)
2. Forsati, R., Mahdavi, M., Shamsfard, M., Meybodi, M.R.: Efficient stochastic algorithms for document clustering. Inf. Sci. **220**, 269–291 (2013)

3. Song, W., Park, S.C.: Genetic algorithm-based text clustering technique. In: Jiao, L., Wang, L., Gao, X., Liu, J., Feng, W. (eds.) Advances in Natural Computation. Lecture Notes in Computer Science, vol. 4221, pp. 779–782. Springer, Berlin (2006)
4. Sreekumar, A., Dhanya, P.M., Jathavedan, M.: Implementation of text clustering using genetic algorithm. Int. J. Comput. Sci. Inf. Technol. **5**(5), 6138 (2014)
5. Zahran, B.M., Kanaan, G.: Text feature selection using particle swarm optimization algorithm. World Appl. Sci. J. **7**, 69–74 (2009)
6. Feng, L., Qiu, M.-H., Wang, Y.-X., Xiang, Q.-L., Yang, Y.-F., Liu, K.: A fast divisive clustering algorithm using an improved discrete particle swarm optimizer. Pattern Recogn. Lett. **31**(11), 1216–1225 (2010)
7. Vaijayanthi, P., Natarajan, A.M., Murugadoss, R.: Ants for document clustering. IJCSI Int. J. Comput. Sci. Issues **9**(2), 493–499 (2012)
8. AbdelHamid, N.M., Halim, M.B.A., Fakhr, M.W.: Bees algorithm-based document clustering. In: International Conference on Information Technology, p. 1 (2013)
9. Hassanzadeh, T., Meybodi, M.R.: A new hybrid approach for data clustering using firefly algorithm and k-means. In: International Symposium on Artificial Intelligence and Signal Processing (AISP), 2012 16th CSI, pp. 007–011 May 2012
10. Mohammed, A.J., Yusof, Y., Husni, H.: Weight-based firefly algorithm for document clustering. In: Herawan, T., Deris, M.M., Abawajy, J. (eds) Proceedings of the First International Conference on Advanced Data and Information Engineering (DaEng-2013), volume 285 of Lecture Notes in Electrical Engineering, pp. 259–266. Springer, Singapore (2014)
11. Rui, T., Fong, S., Yang, X.-S., Deb, S.: Nature-inspired clustering algorithms for web intelligence data. In: International Conferences on Web Intelligence and Intelligent Agent Technology (WI-IAT), 2012 IEEE/WIC/ACM, vol. 3, pp. 147–153 (2012)
12. Mizooji, K.K., Haghighat, A.T., Forsati, R.: Data clustering using bee colony optimization. In: ICCGI 2012, The Seventh International Multi-Conference on Computing in the Global Information Technology, pp. 189–194 (2012)
13. Salton, G., Wong, A., Yang, C.S.: A vector space model for automatic indexing. Commun. ACM **18**(11), 613–620 (1975)
14. Salton, G., Buckley, C.: Term-weighting approaches in automatic text retrieval. In: Information Processing and Management, pp. 513–523 (1988)
15. Yang, X.-S.: Nature-Inspired Metaheuristic Algorithms. Luniver Press (2008)
16. Pham, D.T., Ghanbarzadeh, A., Koc, E., Otri, S., Rahim, S., Zaidi, M.: The Bees Algorithm. Notes by Manufacturing Engineering Centre, Cardiff University, UK (2005)
17. Yang, X.-S.: Firefly algorithms for multimodal optimization. Lect. Notes Comput. Sci. **5792**, 169–178 (2009)
18. Thai thesis database: http://sansarn.com/lexto
19. Lexto: http://sansarn.com/lexto

The Development of Semi-automatic Sentiment Lexicon Construction Tool for Thai Sentiment Analysis

Hutchatai Chanlekha[1]([⊠]), Wanchat Damdoung[1], and Mukda Suktarachan[2]

[1] Department of Computer Engineering, Kasetsart University, Bangkok, Thailand
fenghtc@ku.ac.th, wanchat111@gmail.com
[2] Department of Linguistics, Kasetsart University, Bangkok, Thailand
fhummds@ku.ac.th

Abstract. Sentiment analysis has gained so much interest from many companies and organizations in Thailand. However, there are a few research studies focused on developing Thai sentiment lexicon, which is an important resource for the sentiment analysis. In this work, we developed a web-based automatic Thai lexicon construction tool. Our tool employed a semi-supervised approach for semi-automatically extracting the sentiment lexicon entries. To reduce a negative impact from unreliable parser, we provide simple heuristic rules and mutual information for recognizing sentiment words and its features. The polarity of recognized sentiment words is automatically identified through a bootstrapping process that utilizes a small set of sentiment seeds, the context coherency characteristics, and statistical co-occurrence. In the evaluation, we received quite fair results for lexicon construction task, 76.06 and 75.28 F-Score for hotel review and laptop review, respectively.

Keywords: Sentiment lexicon · Sentiment analysis · Thai language · Lexicon construction tool · Semi-automatic · Semi-supervised learning

1 Introduction

With the rapid growth of user-generated online contents, vast amount of attention has been focused on analyzing user opinions toward products, services or social issues. In Thailand, sentiment analysis or opinion mining has also gained more and more interest from many companies and organizations. Although sentiment lexicon is an important resource for sentiment analysis, there are very few research studies that focused on developing Thai sentiment lexicon or sentiment lexicon construction tools.

To facilitate a domain-dependent Thai sentiment lexicon construction, we extended our previous work [1] and developed a tool in a form of web application. Our proposed method automatically identifies sentiment words and their polarity from relevant corpus, using a small number of seed sentiment words and their polarity. To support the nature of the sentiment words where their polarity usually depends on their context, we designed our lexicon entry as a triple, composed of sentiment words, topic or feature

T. Theeramunkong et al. (eds.), *Advances in Natural Language Processing, Intelligent Informatics and Smart Technology*, Advances in Intelligent Systems

words, and sentiment polarity. With simple heuristic rules and statistical information, candidate pairs of sentiment words and feature are extracted from a corpus. The polarity of these extracted pairs is identified through a bootstrapping process, which takes into account the contextual information, statistical information, and the sentiment coherency characteristic of the context [2]. Utilizing an advantage of semi-supervised approach, this tool can help reducing time and effort in annotating a training corpus. Using our tool, the human developers are only needed for manually verifying the sentiment entries automatically extracted from the corpus.

The remainder of this paper is organized as follows. The related works are presented in the next section. The proposed lexicon construction approach and the development of the tool are described in detail in Sects. 3 and 4, respectively. Experimentation results are demonstrated in Sect. 5. Finally, we conclude our work in Sect. 6.

2 Related Works

The quality of the sentiment lexicons has a significant impact on the performance of the sentiment analysis tasks. A great number of strategies have been proposed for constructing sentiment lexicon, either manually [3, 4] or automatically. For automatic lexicon construction, many works have exploited the linguistic information available in general dictionaries. The most widely used one is WordNet [5]. These works usually utilized WordNet's semantic relations between two Synsets [6, 21] to expand the number of sentiment entries in the lexicon. Even though WordNet has been translated into Thai [7], there are approximately only 53% approved translations. Because of these limited numbers of translations, using Thai WordNet to create sentiment lexicon could affect the coverage and quality of the resulting lexicon.

Incorporating dictionaries as a source for sentiment lexicon construction can help reduce time and effort in developing a training corpus. However, in this approach, the orientation of the sentiment words usually depends solely on the semantic of the words without taking into account the domain or context that could also affect the orientation. To avoid this limitation, many works have focused on extracting a sentiment lexicon from corpus. The advantage of this strategy is that it could reflect the real usage in the domain of interest and can incorporate the effect of the surrounding context on the polarity of the sentiment word. The corpus-based approaches usually exploit the contextual information, especially from conjunction for identifying the sentiment polarity. Hatzivassiloglou and McKeown [8] proposed the idea that the conjunction between adjectives can provide indirect information about the sentiment orientation. Kanayama and Nasukawa [2] extended this idea and proposed a context coherency. Their assumption is that the clauses with the same polarities tend to appear successively in contexts. Utilizing the context coherency, they associate the sentiment candidates with the same polarities as those adjacent clauses which have already been identified the polarities.

Many works proposed semi-supervised learning approach in the form of bootstrapping method to expand the extracted sentiment lexicon. Banea et al. [9] create sentiment lexicon by performing a bootstrapping process on an online dictionary.

Starting with a small set of seed words, sentiment candidates are extracted from a dictionary. These candidates will be selected to be included in final sentiment lexicon based on the LSA similarity score between itself and the seed word. However, depending on a dictionary, the resulting lexicon could have a limited usage when working with domain-specific applications. Moreover, it is unable to reflect the context-dependent nature of the sentiment words. Qui et al. [10] proposed a boot-strapping approach called double propagation, which utilized the syntactic relations between sentiment words and the features that the sentiment words modify. The key idea of their method is that the sentiment words are almost always associated with features. Thus, the features can be identified by considering the syntactic relation with the recognized sentiment words, and then the extracted features can be used for identifying new sentiment words in the same manner.

Even though there are many public-available English sentiment lexicons, such as SentiWordNet [11], Harvard General Inquirer Lexicon [12], MPQA Subjectivity Lexicon [13], there have been very few research studies that focus on developing Thai sentiment lexicon and tools. Haruechaiyasak et al. [14] proposed a strategy for con-structing Thai language resource for opinion mining. To extract the features and polar words, they used the syntactic patterns which were created from tagged-corpus. Although the extraction task can be done automatically, a lot of time and effort is still needed for corpus annotation. There is also an attempt to translate SentiWordNet to 56 languages, including Thai [15]. However, their approach has its own inherit weak-nesses. Apart from neglecting domain and context dependency, the translation process does not consider the possibility that one word can have multiple translations, which may have different sentiment intensity.

3 Semi-supervised Approach for Thai Sentiment Lexicon Construction

3.1 Lexicon Entry

It has been known that the semantic orientation of many sentiment words depends on its context [16]. Based on this observation, many works have developed sentiment lexicons that include the contextual information to maximize the usability [10, 14, 17].

Here, we design the entry of our lexicon to be represented in the form of:

$$< sentiment, feature, polarity >$$

where

  sentiment is the word representing the opinion of the speaker/writer
  feature is the target of the opinion which is expressed by *sentiment* word
  polarity is the semantic orientation of *sentiment* when it co-occurs with *feature*

The following entries are examples of our output lexicon

    <แพง/expensive, ห้องพัก/room, Negative>
    <อร่อย/delicious, อาหาร/food, Positive>

3.2   Corpus Preparation

We employed two corpora for semi-automatic constructing sentiment lexicon. The first corpus is used directly for extracting sentiment words and features. Preprocessing of this corpus includes word segmentation, sentence segmentation, and part-of-speech tagging. We also employed the word formation tool [18] to group a certain type of compound nouns that were segmented to two separated words by our word tokenizer.

The second corpus is used for collecting the word co-occurrence statistics. The size of this corpus must be large enough for the statistics to be reliable, so we require this corpus to be word-segmented only to minimize the burden in preparing corpus.

All conjunctions and negative words will also be marked up in both corpora.

3.3   The Semi-supervised Approach for Thai Lexicon Construction Tool

In this section, we are going to explain the approach that was implemented in our Thai lexicon construction tool in detail. The process overview is shown in Fig. 1.

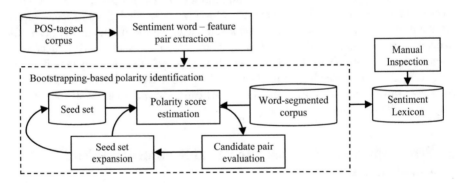

Fig. 1. Process overview

3.3.1   Sentiment and Feature

Sentiment: A vast amount of works on sentiment analysis consider sentiment word to be adjective and adverb. From corpus analysis, we found that, in Thai, when the adjective is used to modify the subject in the sentence without any other verbs, that adjective could be annotated as a verb of the sentence as follows:

"ห้องพัก/room/noun สวย/beautiful/verb"
"โรงแรม/hotel/noun มี/has/verb ห้องพัก/room/noun สวย/beautiful/adjective"

Based on the observation, not only adjective and adverb, the words annotated as verb that has the same surface form as the adjectives in the dictionary will be considered as a candidate sentiment word. Further observation also showed that when an adverb is used to modify an adjectival sentiment, the adverb generally expresses an intensity of the adjective without affecting its orientation. According to this, we will not consider the adverb that modifies adjective as a sentiment word.

Feature: The features are the context of the sentiment word that affects the interpretation of its semantic orientation. From corpus observation, these features are usually the words directly modified by the adjectival or adverbial sentiment, or the subject of the verbal sentiment.

3.3.2 Sentiment—Feature Pair Extraction Methodology

There are two strategies employed in our tool for identifying the sentiment and feature pairs. The first method utilizes the syntactic relations between words. The second method avoids the reliance on syntactic dependency parser by using heuristic rules, which consider only part-of-speech information and word position in the sentence, and statistical information. The details of both methods are described below.

3.3.2.1 Sentiment Word Identification

Based on part-of-speech information, all words that were annotated as adjective or adverb will be recognized as sentiment candidates. For analyzing verb sentiment, we first created a list of adjectives from Thai dictionary. Any words that were annotated as verb and appeared in this list will also be recognized as a sentiment. All these identified sentiment words will act as an anchor for extracting the feature in the following step.

3.3.2.2 Syntactic Rule-Based Sentiment-Feature Pair Recognition

Syntactic relation has been used in many works for identifying sentiment word and feature [2, 14, 17]. In our tool, we use the dependency parser [19] to output the sentences' syntactic dependency structure. From corpus observation, a certain group of relations between sentiment word and other words in the same sentence was selected for extracting sentiment-feature pairs. These dependency relations[1] and the rules for extracting the sentiment-feature pairs are shown below.

Let $sent_{adj}$, $sent_{adv}$, $sent_{verb}$ is a sentiment word with part-of-speech tagged as adjective, adverb, and verb, respectively.

- $mod^*(w, w_{sent})$: Any word w that is modified by the w_{sent}, where w_{sent} is $sent_{adj}$, $sent_{adv}$, or $sent_{verb}$, w will be extracted as a feature of that sentiment word.
- $subj(sent_{verb}, w)$: Any word w that is a subject of the $sent_{verb}$ will be extracted as a feature of that sentiment word.
- $obj(sent_{verb}, w)$: If the $sent_{verb}$ has no subject, any word w that is an object of the $sent_{verb}$ will be extracted as a feature of that sentiment word.

---

[1] Here, relation mod^* is used to represent all types of modifier relation.

3.3.2.3 Statistical-Based Sentiment-Feature Pair Recognition

Syntactic dependency provides a straightforward way for recognizing the sentiment-feature pairs. However, acquiring such information, especially without a reliable syntactic parser, could be problematic. Here, we propose another method that uses a set of simple rules, based on the part-of-speech and relative position between sentiment word and other words in the same sentence, and statistical information to extract candidate sentiment-feature pairs. Brief summary of the rules is explained below.

- $sent_{adj}$ will be paired with all nouns that appear in its front. If there is another verb or sentiment word appears in front of the $sent_{adj}$, only nouns appearing between $sent_{adj}$ and these words will be paired with $sent_{adj}$.
- $sent_{adv}$ will be paired with all verbs that appear in its front. However, if there is an adjective appears in front of the $sent_{adv}$, only verbs appearing between $sent_{adv}$ and the adjective will be paired with $sent_{adv}$.
- $sent_{verb}$ will be paired with all nouns that appear in its front. However, if there are other verbs, relative pronouns, or a specific set of conjunction[2] appearing in front of the $sent_{verb}$, only nouns appearing between $sent_{verb}$ and these words will be paired with $sent_{verb}$.

One of the problems in employing these simple rules is that they tend to over-generate the candidate sentiment-feature pairs. To lessen the problem, the statistical mutual information, as shown in Eq. 1, is used to filter out the candidate pairs.

$$MI(s,f) = \log_2\left(\frac{N \times N_{NEAR(s,f)}}{N_s \times N_f}\right) \qquad (1)$$

where: s is sentiment word, f is feature word, N is the number of words in the corpus, N_w is the number of time the word w appear in the corpus, and $N_{NEAR(s,f)}$ is the number of time that s and f occur less than k words apart from each other, in an order-preserving way. Here, the k is set to 6.

Only the sentiment-feature pairs with the value of mutual information higher than a predefined threshold, T_{recog}, will be kept as a candidate for the sentiment lexicon.

3.3.3 Sentiment—Feature Pair Polarity Identification

3.3.3.1 Conjunction and Negation Processing

Previous research studies [16, 8, 2] have shown the usefulness of the conjunctions in determining the sentiment polarity of a clause. To utilize the context coherency characteristic, we assigned a score to the conjunctions, as well as negative words. Such score reflects how these words affect the polarity of the sentiment word connected to them. The examples are shown in Table 1.

---

[2] These conjunctions are, such as, แต่/but, อย่างไรก็ดี/however, นอกจากนี้/besides, etc.

Table 1. Example of score assignment for conjunction and negation

Type	Word	Score
Conjunction	"แต่" (but), "ถึงแม้ว่า" (despite), "อย่างไรก็ตาม" (however)	−1
	"และ" (and), "ก็" (also)	+1
Negative	"ไม่" (not)	−1
	"ไม่ค่อย" (not so ...)	−0.8

If there is no explicit conjunction between any two sentences, we will add a dummy conjunction with the score +0.8 to represent the sentence transition.

3.3.3.2   Bootstrapping Strategy for Polarity Identification

To identify the polarity of each sentiment-feature pair, we used a bootstrapping process that can iteratively expand the set of the polarity-recognized seed pairs without human intervention. Starting with a seed set, the process identifies the occurrences of the sentiment word and feature in the corpus. The polarity score is then calculated according to the contextual and statistical information. The pairs with a high confidence score will be added to the seed set. The process will continue until the stopping criterion is satisfied. The detail of the methodology is explained below. To avoid any confusion between a reference of sentiment-feature pair in the seed set and the pair's occurrences in the texts, hereafter we will refer to the former as sentiment-feature pair or seed pair, the latter as sentiment occurrence.

3.3.3.2.1   Initial Seed Set

We separated the seed set into positive and negative seed set. Each entry in both seed sets has the same structure as the lexicon entry explained in Sect. 3.1, with the exception that the polarity is expressed by a score: *<sentiment word, feature, polarity score>*.

The sentiment-feature pairs in the initial seed sets are those that have strong, unambiguous, context-independent polarity, such as "ดี/good", "แย่/bad". Since they are context-independent, the feature of these words is represented as "*". The initial score is set to +5 and −5 for positive seeds and negative seeds, respectively.

3.3.3.2.2   Polarity Score Estimation

To estimate the polarity score of the sentiment-feature pair, the polarity score of each entry is calculated based on local and global information. The final score of sentiment-feature pair is an average of these scores.

- Local Polarity Score Calculation

This strategy utilizes a context coherency characteristic of a text. The process consists of 2 main steps. First, every occurrence of the seed pairs is located in the document. Then the score of each occurrence will be assigned according to the following rules.

R1: If the match is found, the seed's score will be assigned to that occurrence.

R2: If the match is found with a negation word in front of the matched sentiment word, then the negative value of the seed's score will be assigned to that occurrence.

After this step, each matched occurrence will be annotated with the sentiment score, as shown in Fig. 2.

Seed pair: ⟨สวย/beautiful, *, 5⟩

Sentence: โรงแรม/hotel/ncn <sent feature="โรงแรม/hotel" score=5>สวย/beautiful/vi</sent> แต่/but/conj ราคา/price/ncn ห้องพัก/room/ncn สูง/high/vi มาก/very/adv

---

Fig. 2. Output example from seed pair matching process

In the second step, newly-assigned polarity score of these occurrences will be disseminated to its neighbor sentiment occurrences. This context-inferred score is calculated according to the following equation:

$$SO_{context}\left(sf_i^m\right) = \sum_{\langle \overline{sf}_k \rangle \in Seed} \frac{1}{1 + |m - n|} \times SO\left(\overline{sf}_k^n\right) \times Score_{conj}\left(sf_i^m, \overline{sf}_k^n\right) \qquad (2)$$

where: *Seed* is a sentiment seed set; sf_i^m is the *i*th occurrence of the sentiment-feature pair $\langle s, f \rangle$ at sentence *m*; $SO(\overline{sf}_k^n)$ is a polarity score of the *k*th occurrence of the sentiment-feature pair $\langle \overline{s}, \overline{f} \rangle$ at sentence *n*. If there is a negation appearing in front of s_k, this score will be multiplied by that negation's score; $Score_{conj}(sf_i^m, \overline{sf}_k^n)$ is a product of the score of the conjunctions that occur between sf_i^m and $\overline{sf}_k^n$.

The first term of the Eq. 2 shows that the impact of neighboring seed decreases in inverse proportion to the distance. The third term of the Eq. 2 reflects an impact of any conjunction appearing between sf_i^m and $\overline{sf}_k^n$, and can be computed according to Eq. 3.

$$Score_{conj}\left(sf_i^m, \overline{sf}_k^n\right) = \prod_{j \in position\,between\,sf_i^m\,and\,\overline{sf}_k^n} weight(conj_j) \qquad (3)$$

where: $weight(conj_j)$ is the conjunction score as described in Sect. 3.3.3.1.

If there is a negation in front of a sentiment occurrence $\langle s_i, f_i \rangle$, its context-inferred score will be multiplied by that negation's score.

The overall local score of a particular sentiment-feature pair will be the average local score of each of its occurrence and can be computed according to Eq. 4.

$$\overline{SO}_{local}(\langle s, f\rangle) = \frac{1}{N} \times \sum_{j=1}^{N} SO_{context}(\langle s_j, f_j\rangle) \tag{4}$$

where: N is the number of occurrences that have a local score value, i.e. $SO_{context} \neq 0$

- Global Polarity Score Calculation

To handle cases that a sentiment word occurs without any other nearby polarity-identified sentiment words, we employ global information, which is a statistical co-occurrence with a positive or negative reference inferred from a large untagged corpus, for calculating a polarity score. The global polarity score is determined by using PMI-IR [20], with the adaptation to accommodate a multiple reference words usage.

$$SO_{PMI-IR}(s, f) = \log_2 \left[\frac{\left(\left(\prod_{p_i \in P}(\text{hits}(\text{NEAR}((s, f), p_i)) + 0.01) \right) \prod_{n_i \in N} \text{hits}(n_i) \right)}{\left(\left(\prod_{n_i \in N}(\text{hits}(\text{NEAR}((s, f), n_i)) + 0.01) \right) \prod_{p_i \in P} \text{hits}(p_i) \right)} \right] \tag{5}$$

where P is a set of positive seed; N is a set of negative seed; $hits(z)$ is the number of hits returned given the query z; $\langle s, f\rangle$ is a candidate sentiment-feature pair; $NEAR(x, y)$ is a co-occurrence of x and y, i.e. x and y occur less than ten words apart from one another.

During this step, we also consider the occurrences of negation or conjunction while counting for hits. In the case that the sentiment occurrence is preceded by negation word, the hit will be counted for the opposite polarity as follows:

"เค้ก/cake/ncn ไม่/not/neg อร่อย/delicious/vi แข็ง/hard/vi มาก/very/adv"
"เค้ก/cake/ncn อร่อย/delicious/vi ไม่/not/neg แข็ง/hard/vi เลย/completely/particle"

Suppose that $\langle$"อร่อย/delicious", $*\rangle$ is a positive seed pair. In both sentences, "แข็ง/ hard" will be counted as a hit for negative hit.

- Overall Polarity Score

The overall polarity score of the sentiment-feature pair is the average of the local score and global score.

$$Score(\langle s, f\rangle) = w_{global}SO_{PMI-IR}(\langle s, f\rangle) + (1 - w_{global})\overline{SO}_{local}(\langle s, f\rangle) \tag{6}$$

3.3.3.2.3   Seed Set Expansion

At the end of each iteration, the seed set will be expanded with the new sentiment-feature pairs. Only the top 3 pairs with the score higher than the threshold T_1 and the bottom 3 pairs with the score lower than the threshold T_2 will be added to the positive seed set and negative seed set respectively. The expanded seed set will then be used to identify other sentiment occurrences' polarity score in the next iteration. The bootstrapping process will be terminated when the polarity of every sentiment-feature pair is recognized, or there is no sentiment-feature pair with the score higher than T_1 or lower than T_2.

3.3.4   Lexicon Expansion with Thesaurus

When working with a small-size or diversity-lack training corpus, one problem that most bootstrapping-based systems usually faced is a low coverage of the output. In order to assist in improving the coverage of the lexicon, in this tool, we provided a lexicon expansion function which can be used for augmenting the output lexicon. In the expansion process, we utilized the semantic relation available in Thai WordNet [7], by following the synonym relations, i.e. synsets, of the features. If there are synonyms of sentiment words, each of these synonyms will be paired with the features of that sentiment word and added to an expanded list for manual verification. This process is also applied with the feature words as well.

Even though this task is available for both sentiment and feature words, it must be performed independently. The reason that we separated the expansion of sentiment words and feature words is to reduce the combinatory explosion, which will cause a burden in a manual verification process.

Although this function could help in increasing the coverage, the expanding results are often improper or invalid in terms of the language usage. Such mistakes still occur because in searching for synonyms, only surface form is considered without performing a word sense disambiguation. Moreover, there is also a restriction of vocabulary usage or collocation that prohibits a certain pairs of words to be used together.

4   Thai Sentiment Lexicon Construction Tool

Based on the approach presented in Sect. 3, we developed a Thai sentiment lexicon construction tool in a form of web application. The tool is shown in Figs. 3, 4.

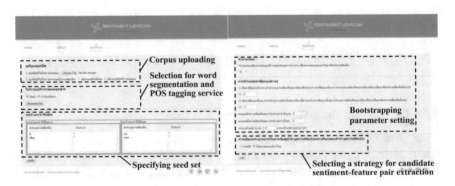

Fig. 3. Tool interface for corpus uploading, parameter setting, and lexicon editing

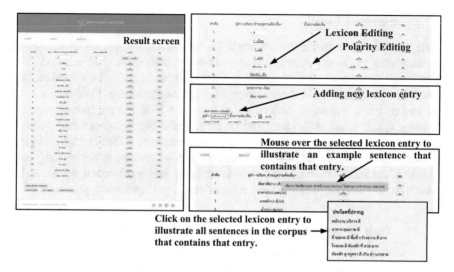

Fig. 4. Tool interface for lexicon entry verification

This tool allows users to upload their own corpus, as well as freely set all parameters mentioned in Sect. 3, including initial seed set. Tools also provide an interface for manipulating the output sentiment lexicon by editing, adding, and deleting the lexicon entry. The users can verify not only the final lexicon, but also the immediate seed set generated during the bootstrapping process by setting the number of iterative batches. This intermediate verification could help in controlling seed quality, and consequently, result in improving the accuracy of the final output. To support the verification process, the users can click on the sentiment-feature entry to view all sentences in the corpus that contain that selected pair. This tool also provides export function that will facilitate the users in exporting the output lexicon in a text format.

5   Experimental Result

5.1   Experiment Settings

To evaluate our proposed approach, we performed the experiments on two corpora from two domains, namely hotel reviews[3] and computer laptop review.[4] There are 100,000 sentences in hotel reviews, with 3000 POS-tagged sentences. For computer laptop review, there are 50,000 sentences. 2000 sentences of which are POS-tagged. For the POS-tagged part in both domains, we manually extracted the sentiment-feature pairs and assigned the polarity. This manually-extracted lexicon will be used as a gold-standard for evaluation.

---

[3] The hotel reviews were collected from Agoda website: http://www.agoda.co.th.

[4] The laptop reviews were collected from Notebookspec website: http://www.notebookspec.com.

5.2   Experimental Results

In this section, we report the experiment on two tasks, the sentiment feature pair extraction and sentiment—feature pair polarity identification. Table 2 shows the experimental results of the sentiment-feature pair extraction. In statistical-based method, we performed several experiments for parameter tuning. However, due to a limit space, we will present only the best results which were achieved when the threshold, T_{recog}, is set to 3.5 for hotel reviews, and 3 for laptop reviews. From the results, while the performances of both methods are comparable in laptop reviews, the performance of syntactic rule-based method in hotel reviews is significantly lower than that of statistical-based method. This is probably due to the casual written style in the hotel review corpus, which causes a lot of errors in dependency parsing.

Table 2. Experimental results of the sentiment-feature pair extraction

Method	Hotel review corpus			Laptop review corpus		
	Precision	Recall	F-score	Precision	Recall	F-score
Syntactic rule-based	65.12	54.60	59.40	79.51	65.57	71.87
Statistical-based	74.33	76.30	75.30	78.88	66.04	71.89

To create a seed set for sentiment-feature polarity identification, we collected high-frequency sentiment words from our training corpus and then manually selected the words which the polarity is independent from the context given the domain to be included in the seed set. From the observation, we found that, in many domains, people usually expressed negative opinion by using negation with positive words, such as "ไม่ดี/not good". This causes the number of sentiment words with negative polarity becoming significantly lower than those with positive polarity. This imbalance characteristic may have a negative impact on the accuracy of polarity identification. To reduce this effect, we tried to extend the number of negative seed set with the terms created by appending negative words to positive seeds, for example "ไม่ดี/not good" "ไม่สวย/not beautiful". Comparing between the polarity identification performed by using the seed set with and without the negative-seed expansion, the experimental results showed that the expanded seed set yielded a better accuracy. In hotel review, the best result achieved when using the seed set without negative-seed expansion is 72.73% F-score compared with 76.06% F-score when using the expanded negative seed set. The result for laptop review shows the same tendency, 75.28 and 64.74% F-score, when using the seed set with and without negative-seed expansion respectively.

   To evaluate an impact of the number of seed words to the polarity identification performance, we also performed an experiment with different numbers of seed words, varying from 2 words each in positive and negative sets, to 4 words each in positive and negative sets. The experiments were performed on the semantic-feature pairs extracted by using the statistical-based method. The seed words used in the experiments are summarized in Table 3. Noted that the negative seed set, represented as $n^{(n)}_{neg}$, in the following experiments was expanded with another set of n seeds forming by

Table 3. Seed words used for sentiment-feature pair polarity identification

Domain	#of seeds	Positive seed	Negative seed
Hotel review	2 pos/2$^{(2)}$ neg	ดี/good, สวย/beautiful	แพง/expensive, แย่/bad
	3 pos/3$^{(3)}$ neg	ดี/good, สวย/beautiful, สะอาด/clean	แพง/expensive, แย่/bad, สกปรก/dirty
	4 pos/4$^{(4)}$ neg	ดี/good, สวย/beautiful,สะอาด/clean, อร่อย/delicious	แพง/expensive, แย่/bad, สกปรก/dirty, เก่า/old
Laptop review	2 pos/2$^{(2)}$ neg	ดี/good, สวย/beautiful	แพง/expensive, แย่/bad
	3 pos/3$^{(3)}$ neg	ดี/good, สวย/beautiful, ถูก/cheap	แพง/expensive, แย่/bad, แข็ง/hard
	4 pos/4$^{(4)}$ neg	ดี/good, สวย/beautiful, ถูก/cheap, แรง/high performance	แพง/expensive, แย่/bad, แข็ง/hard, หนัก/heavy

appending "ไม่/not" to each word in the associated positive set. The best results achieved from different numbers of seed set are shown in Table 4.

Table 4. Experimental results of the sentiment-feature pair polarity identification

Domain	#of seeds	Threshold	Precision	Recall	F-score
Hotel review	2 pos/2$^{(2)}$ neg	$T_1 = 4; T_2 = -3.5$	84.10	69.18	75.91
	3 pos/3$^{(3)}$ neg	$T_1 = 4; T_2 = -3$	84.14	69.40	**76.06**
	4 pos/4$^{(4)}$ neg	$T_1 = 4; T_2 = -3.5$	84.10	69.18	75.91
Laptop review	2 pos/2$^{(2)}$ neg	$T_1 = 4; T_2 = -3$	83.66	67.87	74.94
	3 pos/3$^{(3)}$ neg	$T_1 = 4; T_2 = -3$	84.08	67.87	75.11
	4 pos/4$^{(4)}$ neg	$T_1 = 4; T_2 = -3$	84.50	67.87	**75.28**

The experiments show that the number of seed words has a positive, although not significant, impact on the performance in terms of accuracy. The highest F-score for hotel reviews and laptop reviews are marked with bold-face.

For the sentiment-feature polarity identification, we also performed an experiment to evaluate various threshold settings. As in previous experiments, this experiment was also done on the semantic-feature pairs extracted by using the statistical-based method. Even though a number of parameter settings have been evaluated, in Table 5, we report only the best results achieved for precision, recall, and F-score. The highest precision, recall, and F-score achieved for hotel reviews and laptop reviews are marked with bold-face.

Table 5. Experimental results of the sentiment–feature pair polarity identification

Domain	#of seeds	Threshold	Precision	Recall	F-score
Hotel review	4 pos/4$^{(4)}$ neg	$T_1 > 0; T_2 < 0$	79.12	**71.40**	75.06
	3 pos/3$^{(3)}$ neg	$T_1 = 4; T_2 = -3$	**84.14**	69.40	**76.06**
Laptop review	3 pos/3$^{(3)}$ neg	$T_1 > 0; T_2 < 0$	77.06	**71.49**	74.17
	4 pos/4$^{(4)}$ neg	$T_1 = 4; T_2 = -3$	**84.50**	67.87	**75.28**

Although the quantitative results are quite promising, we found that the variety of the sentiment extracted is not high due to the limited size of POS tagged corpus we used for training.

Since the manual inspection is needed at the end of the construction process to ensure the quality of the lexicon, the users might want to sacrifice the precision to maximize the recall so the tool can output as many sentiment-feature candidates as possible. This could be achieved by adjusting the threshold parameters such as reducing the value of T_{recog} and T_1, and increasing the value of T_2.

6   Conclusion

In this paper, we present the development of the Thai sentiment lexicon construction tool. To reduce time and effort of the human developers in manually creating the lexicon or a training corpus, we employed semi-supervised approach with minimal human intervention. The resulting lexicon from our approach contains not only the sentiment words and polarities, but also the features.

Our approach utilizes sentiment co-occurrence and the context coherency characteristic in propagating the sentiment polarity to an unresolved sentiment-feature pair. The evaluation shows that our approach is fairly reliable in assigning the polarity to the sentiment-feature pair. However, the variety of the sentiment extracted is not high due to the limited size of part-of-speech tagged corpus we used for training. The results also indicate that, with error-prone dependency parser, simple heuristic rules with statistical approach can outperform the syntactic-based approach.

References

1. Damdoung, W., Chanlekha, H., Kawtrakul, A.: A context-induced bootstrapping approach for constructing contextual-dependent Thai sentiment lexicon. In: 10th SNLP, pp. 225–230. Thailand (2013)
2. Kanayama, H., Nasukawa, T.: Fully automatic lexicon expansion for domain-oriented sentiment analysis. In: the 2006 EMNLP, pp. 355–363. Association for Computational Linguistics, Pennsylvania (2006)
3. Hu, M., Liu, B.: Mining and summarizing customer reviews. In: 10th ACM SIGKDD International Conference on Knowledge Discovery and Data Mining, pp. 168–177. ACM, New York (2004)
4. Wilson, T., Wiebe, J., Hoffmann, P.: Recognizing contextual polarity in phrase-level sentiment analysis. In: HLT/EMNLP 2005, pp. 347–354. Association for Computational Linguistics, Pennsylvania (2005)
5. Miller, G.A., Beckwith, R., Fellbaum, C., Gross, D., Miller, K.J.: Introduction to WordNet: an on-line lexical database. Int. J. Lexicogr. 3, 235–244 (1990)
6. Kamps, J., Marx, M., Mokken, R.J., De Rijke, M.: Using WordNet to measure semantic orientations of adjectives. In: 4th LREC, pp. 1115–1118. ELRA (2004)
7. Asian WordNet Project. http://www.asianwordnet.org

8. Hatzivassiloglou, V., McKeown, K.R.: Predicting the semantic orientation of adjectives. In: 35th ACL and 8th EACL, pp. 174–181. Association for Computational Linguistics, Pennsylvania (1997)
9. Banea, C., Mihalcea, R., Wiebe, J.: A bootstrapping method for building subjectivity lexicons for languages with scarce resources. In: 6th LREC, pp. 2764–2767. ELRA (2008)
10. Qiu, G., Liu, B., Bu, J., Chen, C.: Expanding domain sentiment lexicon through double propagation. In: 21st IJCAI, pp. 1199–1204. The AAAI Press, California (2009)
11. Baccianella, S., Esuli, A., Sebastiani, F.: SentiWordNet 3.0: an enhanced lexical resource for sentiment analysis and opinion mining. In: 7th LREC, pp. 2200–2204. ELRA (2010)
12. Stone, P.J., Dunphy, D.C., Smith, M.S.: The General Inquirer: A Computer Approach to Content Analysis. The MIT Press, Massachusetts (1966)
13. Wiebe, J., Wilson, T., Cardie, C.: Annotating expressions of opinions and emotions in language. Language Resour. Eval. **39**, 165–210 (2005)
14. Haruechaiyasak, C., Kongthon, A., Palingoon, P., Sangkeettrakarn, C.: Constructing Thai opinion mining resource: a case study on hotel reviews. In: 8th Workshop on Asian Language Resources, pp. 64–71. CIPS, Beijing (2010)
15. Das, A., Bandyopadhyay, S.: Towards the global SentiWordNet. In: 24th PACLIC, pp. 799–808. Institute for Digital Enhancement of Cognitive Development, Waseda University (2010)
16. Ding, X., Liu, B., Yu, P.S.: A holistic lexicon-based approach to opinion mining. In: the 2008 WSDM, pp. 231–240. ACM, New York (2008)
17. Jijkoun, V., de Rijke, M., Weerkamp, W.: Generating focused topic-specific sentiment lexicons. In: 48th ACL, pp. 585–594. Association for Computational Linguistics, Pennsylvania (2010)
18. Pengphon, N., Kawtrakul, A., Suktarachan, M.: Word formation approach to noun phrase analysis for Thai. In: 5th SNLP, pp. 277–282. Thailand (2002)
19. Sudprasert, S.: Design and development of a lattice structure dependency parser for under-resourced languages. Dissertation, Kasetsart University (2010)
20. Turney, P.D.: Thumbs up or thumbs down?: semantic orientation applied to unsupervised classification of reviews. In: 40th ACL, pp. 417–424. Association for Computational Linguistics, Pennsylvania (2002)
21. Peng, W., Park, D.H.: Generate adjective sentiment dictionary for social media sentiment analysis using constrained nonnegative matrix factorization. In: 5th ICWSM, pp. 273–280. The AAAI Press, California (2011)

The Effect of Automatic Tagging Using Multilingual Metadata for Image Retrieval

Arinori Takahashi and Reiko Hishiyama[✉]

Graduate School of Creative Science and Engineering, Waseda University,
3-4-1 Okubo, Sinjuku-Ku, Tokyo 169-8555, Japan
arinori@ruri.waseda.jp, reiko@waseda.jp

Abstract. One of multimedia content acquisition method on the Web is using tag which is describe of content. However, different words may be tagged to the same or similar subjects, because tagging is performed in an essentially arbitrary manner by a human. In addition, tags are provided in specific language from the content. This model proposes the method of automatically tagging which compensate insufficient described content by using multilanguage. Our experimental evaluations showed that users could obtain more appropriate images with this method.

Keywords: Image retrieval · Automatic tagging · Multilingual retrieval ·
Language grid

1 Introduction

One method of multimedia content acquisition on the Web is to use tags that describe the content. However, different words may be tagged to the same or similar subjects, because tagging is performed in an essentially arbitrary manner by a human. Therefore, if an appropriate tag is not provided, the usual method of directly utilizing tag information is ineffective. In general, a description of the content is only given in one language, which is derived from the content.

Users can make information requests if they feel that they have insufficient knowledge to achieve the objectives of information retrieval [1]. It has been proposed that information requests can be classified into four stages, from Q1–Q4, and that image retrieval using tag information can be based on this classification. If the user state is "Adjusted Request," which suggests insufficient identification (the user can guess the source of the requested information), the user has only performed a retrieval. However, if the user state is Q1, known as an "intuitive request," this is not able to be the language, Q2 which called "consciousness requested" this is not able to linguistic expressions only to ambiguity and Q3 which called "formalized request" this is able to be linguistic expressions, the user is difficult to choice the appropriate keywords and then get the content that does not comply to their intentions.

Therefore, using only search queries or tags based on the user's language, the effective acquisition of content is expected to be difficult. Thus, this study targets an automatic tagging system that uses multiple languages to overcome the problem of

© Springer International Publishing AG 2018
T. Theeramunkong et al. (eds.), *Advances in Natural Language Processing,*
Intelligent Informatics and Smart Technology, Advances in Intelligent Systems

insufficient tag information. The effect on appropriate image acquisition resulting from the proposed system is evaluated experimentally.

2 Related Work

Parton et al. [2] proposed a method for translating text and queries to improve the accuracy of cross-language retrieval. This study did not verify the effect of the combination of cross-language queries. Varshney et al. [3] proposed a query suggestion with corpus method to solve the problem of incorrect results being returned by incomplete dictionaries, incorrect machine translation systems, and unfit translations of the same meaning in different languages when users search for information whose content is described in a language unknown to the user. However, they did not consider metadata such as geographic information obtained from the query. To facilitate effective *image retrieval*, Zakaria et al. [4] tried to enrich content descriptions by combining the collective knowledge generated from georeferenced metadata and user tags without relying on metadata. However, this retrieval method did not consider the user's intentions directly, and multilingual queries were not investigated. Leuken et al. [5] proposed the presentation of various results using image features to assist in the acquisition of appropriate images when the user's request is ambiguous. This does not consider the case where no appropriate search results are obtained. Wu et al. [6] proposed an algorithm for solving an optimization problem that stores appropriate visual information and tag information when automatic tagging is applied to the image. However, although the image features were considered in part of the automatic tagging, this method cannot directly consider the appearance, because image features are not utilized as metadata.

3 Proposed Method

In this study, we propose a method that obtains more appropriate images for the user. First, image retrieval is performed to meet the user's requirements. To utilize this more effectively, a wide range of images must be retrieved to ensure an appropriate selection for the user. Therefore, increasing the number of candidate images is an effective approach for returning more appropriate images. To increase the number of candidates, we aim to enrich the content description by utilizing metadata in the images for a set of searched content that is often poorly described. Thus, we improve the retrieval precision by increasing the number of tag candidates to be searched, while filling the description of the content. Figure 1 shows the flow of the proposed method.

This method may be beneficial for tourists who are searching for sightseeing information. First-time visitors are more interested in general sightseeing than repeat visitors [7], whereas the latter want to see unusual and unfamiliar sights. In addition, many European repeat visitors may search for more unusual parts of London [8]. Thus, they expect more specific sights. For example, the image might be a different color for a limited time, or could have been taken from another position. Tourists can search for

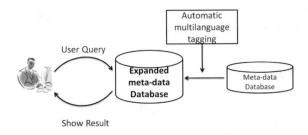

Fig. 1. Overview of proposed method

specific images in their native language, and can then imagine what they will see when they go sightseeing after viewing various images.

The proposed system comprises automatic tagging, language unification, and color extraction, as shown in Fig. 2.

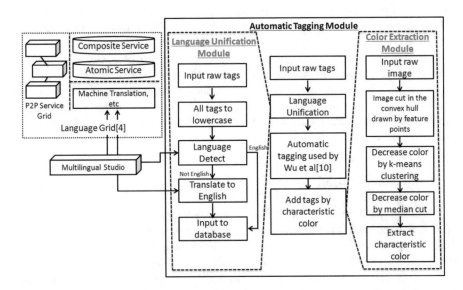

Fig. 2. Details of proposed method

Flow of the Proposed Method

The automatic tagging module first receives image tags. Raw tags are then unified across languages by the language unification module. Next, the automatic tagging approach developed by Wu et al. [6] is used to unify the tags across languages. After this, the characteristic color of the image is received from the color extraction module as the tags are provided to the images. The system then assigns tags to images considering the extracted characteristic color. Finally, a metadata search is conducted, and the results are displayed to the user. The number of images to be displayed at this time is set to 30, because this number can be adequately displayed on typical screens.

Automatic Tagging Module

The automatic tagging module adds processes before and after the previous method used by Wu et al. [6]. As a pre-process, this module unifies the language of the tags by passing the tag set to the language unification module. This is because the values of words that have the same meaning but different characters may not be properly evaluated by their distributed value. For example, "red" could be expressed in a tag as "red," "rouge," "rot," "rojo," or "赤." Therefore, we prevent the values becoming dispersed by unifying the language in advance. As a post-process, we apply tags to images after evaluating the raw tags using the automatic tagging method. At this time, the system detects the characteristic color by passing the image for processing to the color extraction module. This module determines whether to assign the tags or not based on the detected characteristic color. This is because color information is often improperly evaluated because of noise, even though the landmark color is very important information used in the search process. Therefore, tags are assigned in consideration of color information by analyzing the image after the tag evaluation step.

Language Unification Module

The language unification module unifies the language of the raw tags. It discriminates languages through the Language Grid [9]. More than 120 language services are now registered in the Language Grid (at Kyoto University and NECTEC in Thailand), where various atomic services and composite services, such as translation, bilingual dictionary, parallel text, morphological analysis, and text-to-speech, are shared [9]. If translation is required, the language is unified by translation through Language Grid. The detailed process is as follows.

First, the module converts all tags to lowercase, in an attempt to reduce the duplication of tags. This is because a lot of duplication is caused by slight differences in the raw tags. Second, this module investigates the language of the tag using the language identification service of Language Grid [9]. At this point, if the language of the tag is detected to be anything other than English, the tag is translated to English through the translation service. Raw tags are converted to English, and are then returned to the automatic tagging module. We use English as the unified language because the accuracy of translation and detection is higher, mainly because English is very well used.

Color Extraction Module

The color extraction module extracts the characteristic color of the input image. As a pre-process, this module detects the input image's feature points using ORB [10]. After this, the image is split according to the convex hull drawn by the detected feature points. This is because the background is determined to be noise with respect to the characteristic color in landmark retrieval, as users are interested in the landmark itself rather than the background. The number of colors is then decreased by "k-means clustering" and the "median cut" method. As a result, we obtain three characteristic colors. Two color is the detected order from the direction of the upper. The third is mode in plurality color detected. The seven colors typically classified are black, white, red, yellow, green, blue, and purple. An example is shown in Fig. 3, and the extracted color list is given in Table 1.

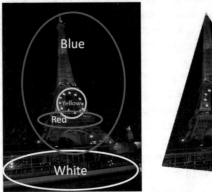 

Figure: Original Image   Figure: Cut Image

Fig. 3. Example of color information extraction

Table 1. Results of each color detection method

	First color	Second color	Mode color
k-means Clustering	Blue	White	White
k-means Clustering + Cut	Blue	Black	White
MedianCut	White	Yellow	Blue
MedianCut + Cut	Black	Blue	Blue
k-means Clustering + MedianCut + Cut	Yellow	Blue	Red

4   Experiments

Four evaluation experiments were conducted. The first is a comparison experiment that did not apply automatic tagging (Experiment 1). The second is a comparison experiment after the automatic tagging process by translating the tag using geographical information (Experiment 2). The third is a comparison experiment that used the automatic tagging method of Wu et al. [6] (Experiment 3). The fourth is a comparison experiment after adding pre/post-processing to Experiment 3, as in the proposed method (Experiment 4).

4.1   Experimental Setup

The experimental targets comprised two landmarks: the Eiffel Tower and the Allianz Arena. The detailed conditions of the Eiffel Tower were "an upward viewing angle, red, and the Eiffel Tower," and 785 images with Eiffel Tower-related tags were retrieved from Flickr. The detailed conditions of the Allianz Arena were "show the logo, red, and Allianz Arena," from which 701 images related to Allianz Arena tags were obtained from Flickr. In addition, a database was constructed with this information, and the tags attached to the images were acquired.

The selection criteria for the retrieval targets were two or more color types in the target and a large number of target images. For example, the Leaning Tower of Pisa has a large number of images, but few color types; Sydney Opera House has many color types, but few images. The queries were specified as "Location Name + Color." If the query included the terms "lit up" or "illuminated," this had no effect on the retrieval accuracy, because images labeled as "red" included the images with "lit up." Thus, the search query was composed as "name of place + color."

5 Results and Discussion

5.1 Results

The precision, recall, and F-measure in each experiment are presented in Tables 2, 3, 4, and 5. Examples of the experimental results obtained are shown in Figs. 4, 5, 6, 7, 8, 9, 10, and 11. In addition, the images obtained in the search results were counted and classified into four classes: "landmark different", "landmark met", "query met", and "positive image". This classification criteria is based on divided query. "landmark different" is nothing to match user query. "landmark met" is half matched user query. "query met" is matched user query but it's not appropriate for user. The results are shown in Figs. 12 and 13.

Table 2. Results of experiment 1

	Eiffel Tower (Ex1-1)	Allianz Arena (Ex1-2)
Precision	0.33	0.50
Recall	0.12	0.17
F-measure	0.18	0.25

Table 3. Results of experiment 2

	Eiffel Tower (Ex2-1)	Allianz Arena (Ex2-2)
Precision	0.37	0.40
Recall	0.13	0.13
F-measure	0.20	0.19

Table 4. Results of experiment 3

	Eiffel Tower (Ex3-1)	Allianz Arena (Ex3-2)
Precision	0.27	0.30
Recall	0.10	0.10
F-measure	0.14	0.15

Table 5. Results of experiment 4

	Eiffel Tower (Ex4-1)	Allianz Arena (Ex4-2)
Precision	0.43	0.50
Recall	0.16	0.17
F-measure	0.23	0.25

Fig. 4. Experiment 1 results for Eiffel Tower

Fig. 5. Experiment 1 results for Allianz Arena

5.2 Discussion

Successful Case Figs. 12 and 13 show that "positive image" and "query met" increase after automatic tagging, while the "landmark different" number decreases.

Additionally, Fig. 6 shows that automatic tagging using geographic information detects noisy images in the top results, because it does not have a noise-elimination function. Moreover, Figs. 9 and 11 show a decreasing number of noisy images towards

Fig. 6. Experiment 2 results for Eiffel Tower

Fig. 7. Experiment 2 results for Allianz Arena

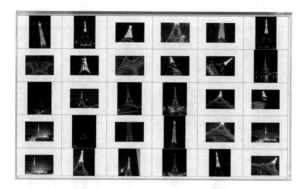

Fig. 8. Experiment 3 results for Eiffel Tower

the bottom of the results. Therefore, by simply extending the tag, we can increase the number of positive images, as shown in Fig. 12. However, if there is a lot of noise, this method is insufficient.

Fig. 9. Experiment 3 results for Allianz Arena

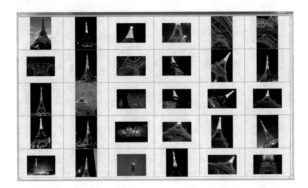

Fig. 10. Experiment 4 results for Eiffel Tower

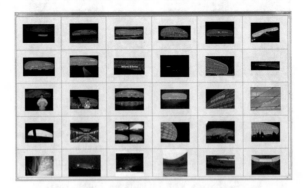

Fig. 11. Experiment 4 results for Allianz Arena

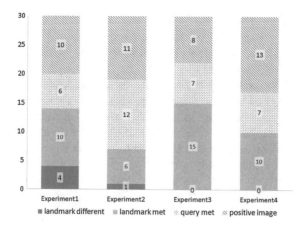

Fig. 12. Comparison of results for Eiffel Tower

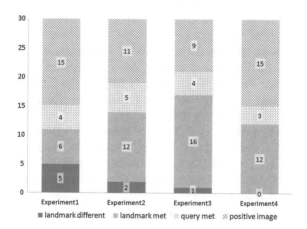

Fig. 13. Comparison of results for Allianz Arena

Tables 4 and 5 show that adding pre/post-processing increases the precision rate more than other methods when performing automatic tagging. There are also many Eiffel Tower images whose primary color is blue, which is clearly different from the user query in the previous automatic tagging method in Fig. 8. In addition, the assigned tag specifically indicates "red." On the other hand, the proposed method assigns tags in consideration of characteristic color, so Fig. 10 does not have a completely different color image.

Figures 5 and 7 show many white Allianz Arena images, which do not match the search query, but Fig. 11 shows little presence of such different images.

These results suggest that considering the metadata in automatic tagging is an effective means of increasing the precision of search results and meeting more of a user's verbalized requests, even if indirectly reflected.

Challenging Case The middle image in the second column of Fig. 10 is mainly blue, although the red specified by the search query is one of the sub-colors. Additionally, images with red and blue stripes affixed to four Allianz Arena images are difficult to determine (see Fig. 11).

The proposed method does not consider these composite colors. Accordingly, simply assigning tags of "red" or "blue" is insufficient, so there is a need for a system that can consider composite colors such as "red and blue." In addition, the Eiffel Tower can be considered as either red, blue, or red and blue. However, the proposed method cannot determine which state is permitted by the user. Therefore, a system that reflects the user's unverbalized requests is needed.

Finally, this system cannot distinguish more detailed requests in matching a landmark. For example, it is not possible to distinguish a "complete picture and steel parts of tower" or "appearance and interior." Therefore, it is necessary to formulate information that considers the contents of an image.

6   Conclusion and Future Work

In the proposed system, more appropriate images are returned by our automatic tagging, which effectively utilizes the metadata and information contained within an image. This study contributes to improving the search accuracy by assigning more appropriate tags in the image database. After identifying the characteristic color of an image, it was possible to readily obtain positive images via automatic tagging by extracting the color of an image as metadata.

In future research, evaluations will be performed with a wide range of images to consider cases with test subjects. To further increase the precision of the proposed method, we intend to introduce ideas that facilitate tagging and more accurate automatic tagging, an automatic correction system for mistakenly assigned tags, the provision of appropriate information for complex states, supplementary information for portions that users cannot verbalize, and the inclusion of detailed information so as not to inhibit the feasibility of searches.

Acknowledgements. This work was supported by a Grant-in-Aid for Scientific Research (S) (24220002, 2012–2016) from Japan Society for the Promotion of Science (JSPS).

References

1. Taylor, R.S.: Question-negotiation and information seeking in libraries. Assoc. Coll. Res. Libraries **29**, 178–194 (1968)
2. Parton, K., McKeown, K.R., Allan, J., Henestroza, E.: Simultaneous multilingual search for translingual information retrieval. In: Proceedings of the 17th ACM conference on Information and knowledge management, pp. 719–728 (2008)
3. Varshney, S., Bajpai, J.: Improving performance of English-Hindi cross language information retrieval using transliteration of query terms **2**(6), 53–59 (2014)

4. Zakaria, L.Q., Hall, W., Lewis, P.: Modelling image semantic descriptions from web 2.0 documents using a hybrid approach. In: Proceedings of the 11th International Conference on Information Integration and Web-based Applications & Services, pp. 306–312 (2009)
5. van Leuken, R.H., Garcia, L., Olivares, X., van Zwol, R.: Visual diversification of image search results. In: Proceedings of the 18th international conference on World wide web, pp. 341–350 (2009)
6. Wu, L., Jin, R., Jain, A.K.: Tag completion for image retrieval. IEEE Trans. Pattern Anal. Mach. Intell. **35**(3), 716–727 (2013)
7. Li, W., Duan, L., Xu, D., Tsang, I.W.H.: Text-based image retrieval using progressive multi-instance learning. In: IEEE International Conference on Computer Vision (ICCV2011), pp. 2049–2055 (2011)
8. Smith, A.: Using major events to promote peripheral urban areas: Deptford and the 2007 Tour de France. Int. Perspect. Festivals Events Paradigms Anal. 3–19 (2009)
9. Murakami, Y., Lin, D., Tanaka, M., Nakaguchi, T., Ishida, T.: Language service management with the language grid. In: The International Conference on Language Resources and Evaluation, pp. 3526–3531 (2010)
10. Rublee, E., Rabaud, V., Konolige, K., Bradski, G.: ORB: an efficient alternative to SIFT or SURF. In: IEEE International Conference on Computer Vision(ICCV2011), pp. 2564–2571 (2011)

The NECTEC 2015 Thai Open-Domain Automatic Speech Recognition System

Chai Wutiwiwatchai$^{(\boxtimes)}$, Vataya Chunwijitra, Sila Chunwijitra,
Phuttapong Sertsi, Sawit Kasuriya, Patcharika Chootrakool,
and Kwanchiva Thangthai

National Electronics and Computer Technology Center, National Science
and Technology Development Agency, Pathumthani, Thailand
{chai.wutiwiwatchai,vataya.chunwijitra,sila.
chunwijitra,phuttapong.sertsi,sawit.kasuriya,
patcharika.chootrakool,kwanchiva.thangthai}
@nectec.or.th

Abstract. We describe the recent development of the NECTEC Thai open-domain automatic speech recognition system. Some of the techniques that were found beneficial over its baseline system are: hybrid word-subword language modeling to enhance the vocabulary coverage in a constraint resource; multi-conditioned noisy acoustic modeling to improve the system robustness using a newly developed large social media speech database; recent state-of-the-art speech features; and lastly, online decoding and speech compression to reduce the processing and data transmission time. These techniques result in a 32.4% word error rate on open-domain noisy speech test sets which is 35.7% relatively lower than the baseline system. The overall system operates in an average 1.2xRT which is promising for real applications.

Keywords: Open-domain · Speech recognition · Thai language

1 Introduction

Large vocabulary continuous speech recognition (LVCSR), automatic speech recognition (ASR) for natural speech with a large lexicon, has gained significant advance both in the research and implementation aspects in the past few years. Several well-known IT companies as well as research institutes have shown their systems running for open-domain speech input. Two recent breakthroughs causing the shift of the technology are big data, the benefit of very large scale data obtained directly from social usage; and deep learning, a newly discovered learning machine capable to capture the high variation of the data. The IBM 2015 system [1] was presented to achieve 23% word recognition error rate on a natural conversational task over telephones. The system took several sophisticated neural-network based algorithms as well as huge 2000-hours training data into account. Google English Voice Search [2] reported its performance of less than 20% word recognition error rate with 230 billion words language modeling data and more than 5000 h acoustic modeling data. The RWTH LVCSR system presented recently its performance on different languages

T. Theeramunkong et al. (eds.), *Advances in Natural Language Processing,*
Intelligent Informatics and Smart Technology, Advances in Intelligent Systems

including Polish, Portuguese, English, and Arabic. Significant improvements were obtained using many modern proposed techniques which made the recognition error rate from over 30% down to less than 20% [3]. These reports convincingly express the applicability of the technology in the near future.

Research on Thai LVCSR in National Electronics and Computer Technology Center (NECTEC) has been conducted since 2003 with a series of publications on Thai continuous speech corpora [4–7]. A report on the system development utilizing the LOTUS corpus showed the first baseline performance at 24.4% word error rate (WER) on a quiet environment, read speech test set [8]. NECTEC has joined Universal Speech Translation Advance Research (USTAR) consortium [9] since 2007 and published a report on Thai ASR for a network-based speech translation service in travel and sport domains [10].

Research and development aiming at open-domain, i.e. ASR with unlimited domains, has just been focused in NECTEC since 2012. Key improvements have been reported consecutively on both the algorithm for reducing out-of-vocabulary (OOV) words and the system architecture suited for service implementation [11–13]. This paper summarizes the key improvements so far integrated in the 2015 NECTEC open-domain ASR system. Comparative experiments over a baseline system along the past years regarding important issues we found on building open-domain ASR and engineering the system are given.

This paper is organized as follows. The next section describes our baseline system built around the year 2012–2013. Section 3 presents improvement issues: hybrid word-subword language modeling, robust acoustic modeling, and run-time system design, respectively. Section 4 shows experiments, Sect. 5 discusses on existing problems and future work, and concludes this paper.

2 Baseline System Development

2.1 Structure of ASR

Our current Thai ASR system has been developed from an open-source KALDI toolkit [14], which is based on weighted finite-state transducer (WFST). Speech parameters given by the feature extraction module are input to the decoding module which takes into account three major components, an acoustic model, a language model, and a pronunciation lexicon. To recognize a speech input, the decoding module constructs a word graph whose word links are tagged with their corresponding language model probabilities. Each word in the graph is expanded to phones according to the given pronunciation lexicon. Each phone refers to its corresponding acoustic model. The speech input travels into the word graph producing potential word paths with top cumulative probabilities. A word recognition result is finally the word path having the highest cumulative probability. In the KALDI toolkit, the acoustic model, the language model, and the pronunciation lexicon are all crafted as WFSTs. An additional context-dependent phone WFST is needed for building a context-dependent acoustic model. All these WFSTs are composited in prior to form a single big WFST used in decoding.

To achieve open-domain ASR, the pronunciation lexicon as well as the language model have to largely cover words used in the language. While the larger the lexicon, the lower the recognition performance, a major problem of making the recognition domain opened becomes how to trade off among the lexicon size and the recognition accuracy. Open-domain ASR also implies the system capability to accept a variety of speech input from different situations, equipments, and environments. Therefore, the system robustness is also another important issue to solve. Last but not least in our open-domain ASR work, an overall system response time has also been taken into account as we aim finally to make the system usable in real applications.

2.2 Development Resources

Table 1 summarizes resources used to develop our baseline Thai open-domain ASR system. The variety of training corpora insists the openness of acceptable speech input over speakers, speaking styles and domains, microphone equipments, and environments. The baseline system utilized in total 224 h of speech for acoustic modeling and 66.74 million-words text for language modeling.

Table 1. Summary of resources used to develop the baseline ASR system

ASR component	Tool	Corpus	Detail
Acoustic model	KALDI	LOTUS	48 speakers, 55 h of article read speech
		LOTUS-BN	147 h of broadcast news speech
		USTAR	22 h of the USTAR speech translation application over smart phone data channels
Language model	SRILM	BEST	7.17 million words from 12 domains
		Thai BTEC	0.83 million words in travel domain
		Thai HIT	0.60 million words in sport domain
		PANTIP	57.32 million words from 8 weblog domains
		LOTUS-BN	0.83 million words of broadcast news

2.3 Building the Baseline System

In acoustic modeling, conventional 13-order Mel-frequency cepstral coefficients (MFCC), their derivatives and second derivatives were extracted from all the speech data presented in the Table 1. Using the KALDI toolkit, the speech features were used to train context-dependent triphone Hidden Markov Models (HMM) covering 75 phones in Thai [4] plus a silence. N-gram language models with Chen and Goodman's modified Kneser-Ney discounting were constructed from the overall text data presented in the Table 1 using the SRILM toolkit [15]. The number of unique words appeared in the training text was more than 140,000 which, for the baseline system, was simply included in the system pronunciation lexicon. As mentioned earlier in the KALDI platform, all these system components were constructed as WFST and recognition can thus use the WFST composition operation. Incorporate the 4-gram language model could be achieved by language model rescoring. The baseline system run-time architecture was simply designed. The ASR server stores input speech in a buffer and starts recognition when the input is completely received. Decoding starts after speech features are extracted from the overall speech. And the output text is returned to the client after the recognition process ends.

3 Key Improvement Issues

Evaluations of the baseline system described above have shown limitations in many issues. In the past three years, many solutions to improve the system have been experimented. This section expresses three key issues we have attacked.

3.1 Hybrid Word-Subword Language Modeling

One of the most important issue to open-domain ASR is the vocabulary coverage. Similar to other languages, new words in Thai have always been invented. Some of them are proper names, person names, and words in social networks. It is hence almost impossible to include all possible words in the system lexicon. Subword unit is one commonly used technique when modeling out-of-vocabulary (OOV) words in many languages as multiple subword units can be combined to form a new word which is not seen before in the training data. Morpheme, a smallest meaningful unit in a language, becomes a natural choice for subword unit especially for highly inflected languages [16, 17]. In Thai, *pseudo-morpheme* (PM), a syllable-like unit in a written form, has been proposed as a subword unit. The definition of the PM is a syllable and if any syllable cannot be represented by a bounded chunk of written text, that PM will span to cope with the least number of syllables whose written text can be bounded. Table 2 shows examples of Thai words and their corresponding PM segments. According to Thai writing rules, PM is more deterministic when compared with word and has been shown to help mitigate the word segmentation problem [18]. Given a word or a string of text, PMs can be determined quite accurately with an automatic syllable segmentation tool [19].

Nevertheless, in recognition, small units usually suffer from acoustic confusability and also cover shorter span of context in an n-gram language model. To resolve these problems, a hybrid language model which combines both unit types, PM and word, has been proposed [11]. In training, the text data were firstly segmented into words. Frequently appeared words were kept in the text and stored in a system lexicon, and the rest words were further segmented into PMs. Only frequently occurred PMs were kept in the lexicon and the rest PMs in the text were marked as unknown units. Then the mixed-unit training text were used to train the N-gram model as usual. While the most frequently used words could be covered by the lexicon, unseen words could also been modeled by sequences of PM units. By this way, we can fully manage the size of the lexicon while keeping the OOV rate minimal.

Table 2. Samples of Thai words and their corresponding pseudo-morphonemes (PM)

Text	Meaning	Word	PM	Pronunciation
กระทรวงวัฒนธรรม	Ministry of culture	กระทรวงวัฒนธรรม	กระ|ทรวง|วัฒน|ธรรม	k r a | s u:a ng | w a t th a n a | th a m
ราชวงศ์จักรี	Chakkri dynasty	ราชวงศ์จักรี	ราช|วงศ์|จัก|รี	r a: t ch a | w o ng | c a k kr i:

3.2 Robust Acoustic Modeling

Although the acoustic model in our baseline system has been built from speech corpora from various speaking environments, its performance against noisy speech is still low. A major cause is that we have not yet directly taken the noisy speech data into training. Besides the data, more advanced speech features and training algorithms have been proposed and have not yet integrated in our system.

Instead of using noise-added speech data, a new speech corpus, LOTUS-SOC [7], has been developed to tackle this problem. The corpus was designed to cover quiet rooms and other 6 noisy environments including cafeteria, busy streets, cars or buses, sky or subway trains, shopping malls, and fast-food restaurants. Nearly 200 speakers were requested to naturally utter the scripts selected from Twitter to mimic the spoken style. Recording was done via a smart phone application created specifically for corpus collection. This database gains an approximately 8.24 SNR ratio in average. An improved acoustic model was constructed by multi-conditioned training, which carefully mixed noisy and clean speech training data.

Many advanced speech features have been proposed recently. In our improved system, MFCCs of a focused speech frame and its 3 surrounding frames, 91 coefficients in total, were collected. Linear Discriminant Analysis (LDA) [20] was applied to reduce the features to 40. Maximum Likelihood Linear Transformation (MLLT) [21] was then used to de-correlate among the 40-order coefficients. This technique (LDA-MLLT) has been widely used and also available in the KALDI toolkit. Discriminative training based on Maximum Mutual Information (MMI) [22] or Maximum Phone Error (MPE) [23] has been a state-of-the-art training algorithm recently as its ability to discriminate ambiguous phones often mistaken in the baseline training algorithm. These discriminative training methods have also been comparatively tested in our system.

3.3 Run-Time System Architecture

The baseline system architecture has to be improved when operating as a real service. Four features have been integrated as illustrated in Fig. 1. First, a Voice Activity Detection (VAD) module was used to segment an input speech at the client side so that the client can gradually send small speech chunks to the ASR server during segmenting. A more complicated VAD module was integrated also in the server side to improve the system robustness against background noise. Second, Speex, an open-source speech codec [24], was incorporated to encode the speech chunk at the client side before data transmission. Speex not only compresses the data, it also suppresses background noise. Since the Speex is a lossy compression, the ASR acoustic model has to be rebuilt from Speex decoded speech to eliminate the mismatch among training and run-time data. Third, data streaming was introduced among client/server transmission to reduce the waiting time required for input buffering. Using the released KALDI online decoder, all ASR models are preloaded before providing a service. The language model rescoring part described in the Sect. 2.3 is skipped to allow real-time processing. This of course degrades the overall system accuracy but tradeoffs for a better real-time factor. Almost all modules in the run-time architecture were developed in a multi-thread concept. Therefore, all the modules can function simultaneously to save the overall processing time. Lastly, our system has been prepared for scalability by introducing a load balance service that can receive multiple speech inputs and distribute to available multiple ASR engines running in different machines.

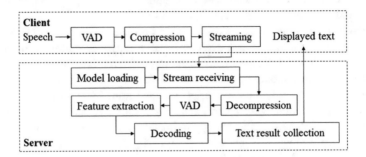

Fig. 1. An improved run-time system architecture

4   Experiments

In this section, experiments aiming to show the performance improvement obtained by using each of the techniques described above. Section 4.1 summarizes the experiments on hybrid word-subword language modeling, Sect. 4.2 on robust acoustic modeling including both recent speech features, multi-conditioned and discriminative training, Sect. 4.3 on the recent system architecture.

4.1   Experiments on Hybrid Word-Subword Language Modeling

To express the effectiveness of hybrid language modeling, a test set was obtained from three subsets: 2200 utterances of 10 speakers from the LOTUS-BN, 300 utterances recorded by 3 speakers in office environment covering 5 genres (newspaper, law, novel, social media and web board), and 2000 utterances from the U-STAR speech translation mobile application.

Figure 2 illustrates comparative results between the hybrid language model system and the baseline system. It is clear that at a much smaller size of the system lexicon, the proposed hybrid technique can even lower the OOV rate, reduce the test set perplexity (which means easier recognition), and preserve the overall PM recognition error. Having this proposed technique, we later included more training text and our final run-time system contains 59,835 lexical units in which 11,035 are PMs.

Table 3 shows examples of recognition results of the baseline system against the hybrid LM system. Words and PMs are separated by "-" and "|", respectively. Words in the Table are OOV words found in the test sets. The first and second rows are proper names while the last row is a Thai transliterated word of the word "alliance". Named-entities and transliterated words are known to be the main causes of OOV. By modeling an OOV word with a sequence of PMs, these OOV words could be correctly recognized by our hybrid LM.

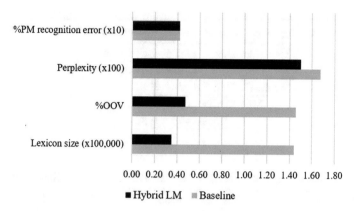

Fig. 2. Experimental results of the hybrid LM system against the baseline system

Table 3. Samples of recognition results of baseline and hybrid LM

Word	Baseline LM	Hybrid LM
บางบุนทอง /b a: ng kh u n th @: ng/	บ้านคลอง /b a: n khl @: ng/	บาง | บุน | ทอง /b a: ng | kh u n | th @: ng/
คลองท่อม /khl @: ng th @ m/	บอง - ทอม /kh @: ng - th @ m/	คลอง | ท่อม /khl @: ng | th @ m/
อัลลายแอนซ์ /z a n l a: j z x n/	อะไร - อัน /z a r a j - z a n/	อัล | ลาย | แอนซ์ /z a n | l a: j | z x n/

4.2 Experiments on Robust Acoustic Modeling

To evaluate the system robustness against noisy speech, another test set was constructed. It contained 3140 utterances from 3 speakers in the LOTUS-BN; 1916 utterances from the U-STAR speech translation mobile application, and 5586 utterances from 14 speakers in 7 noisy environments taken from the LOTUS-SOC.

Table 4 shows evaluation results of the baseline system and improved systems using the LDA-MLLT features, the two discriminative training methods (MPE and MMI), and the multi-conditioned noisy speech training, described in the Sect. 3.2. LDA-MLLT and MMI discriminative training slightly help reducing the Word Error Rate (WER). Multi-conditioned training clearly shows its effectiveness to noise robust. The best setting is the LDA-MLLT, MMI, and multi-conditioned trained system which achieved a 32.4% WER on this open-domain noisy test set, which is 35.7% relatively lower than the baseline WER result.

Table 5 shows examples of recognition results of baseline against multiconditioned training system in the same utterance where words are separated by "-". From the results, we can clearly see that both utterances cannot be correctly recognized by the baseline system, only clean speech in training data, as these speech are recorded under the noisy environments. When using multi-conditioned training, mixing noisy and clean speech, the utterance recognition could be recovered. The multi-conditioned

Table 4. Word error rate (%) results of the systems using LDA-MLLT, discriminative and multiconditioned training compared to the baseline system

Speech features		Training condition	
		Normal	Multi-conditioned
Baseline		50.4	34.9
LDA-MLLT	MPE	63.8	43.3
	MMI	49.1	32.4

Table 5. Samples of recognition results of the baseline and multiconditioned trained system

Utterance	Baseline (Normal)	Multi-conditioned (LDA-MLLT + MMI)
ทั้งที - จริง - ก็ - ไม่ - ระแวง /th a ng th i: - c i ng - k q: - m a j - r a w x: ng/	ทั้งที - จริง - ก็ - ไม่ - เข้าใจดี /th a ng th i: - c i ng - k q: - m a j - kh a w c a j d i:/	ทั้งที - จริง - ก็ - ไม่ - ระแวง /th a ng th i: - c i ng - k q: - m a j - r a w x: ng/
ความ - เสียใจ - มัน - ก็ - เหมือน - ความ - สุข /khw a: m - s i:a j c a j - m a n - k @: - m v:a n - khw a: m - s u k/	ฉัน - เสียใจ - มัน - อยู่ - กับ - มัน - ใช้ได้ /ch a: n - s i:a j c a j - m a n - j u: - k a p - m a n - ch a j d a: j/	ความ - เสียใจ - มัน - ก็ - เหมือน - ความ - สุข /khw a: m - s i:a j c a j - m a n - k @: - m v:a n - khw a: m - s u k/

trained system can be used to recognize noisy speech from various speaking environments.

4.3 Experiments on the Run-Time Performance

According to the newly designed run-time system explained in the Sect. 3.3, we expect the system could operate in a much lower Real-time Factor (RTF); calculated by the total time required from the start of voice recording until the text output is shown, divided by the length of speech input. Since our ASR engine has been designed for scalability, its service can be multiplied and run in parallel to serve concurrent requests. We also expect our system to be acceptably fast in both broadband (WiFi) and narrow-band (3G) conditions. We simulated run-time usage by varying the number of concurrent inputs, the number of available ASR services, and a network condition. The broadband is set 30 Mbps uploading and 30 Mbps downloading, whereas the narrow-band is set 500 Kbps uploading and 1 Mbps downloading.

Figure 3 presents RTF results from both the baseline architecture (denoted as "B" in the graph) and the improved architecture in the Fig. 1 (denoted as "I" in the graph). The results obviously show that at only one ASR service in the narrow-band condition, the new architecture can reduce the RTF down from 3.0xRT to about 1.2xRT, closed to that produced in the broadband network. One run-time service supports up to 5 concurrent requests at about 3.5xRT response time. The system footprint per service is

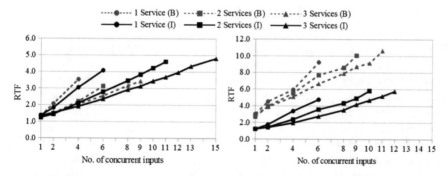

Fig. 3. RTF results of the systems running in a broadband (WiFi) condition (left), and a narrow-band (3G) condition (right), B and I denote the baseline and the improved systems

recommended at least 6.2 GB RAM, 3 GB HDD, 2.6 GHz CPU, and 100 Mbps speed network adapter.

To improve run-time performance, in the system, speex codec was utilized for reducing data size between server and client. Parameters required to be properly set are Constant Bit-Rate (*CBR*), Variable Bit-Rate (*VBR*), Average Bit-Rate (*ABR*), and Quality mode. These parameters are compared as shown in Fig. 4. Q10 denotes the highest sound quality usually set for professional audio work (at 24,600 bps), and Q8 denotes the good sound quality normally set for human voice (at 15,000 bps). According to the Figure, all settings can drastically reduce the size of speech data but we found that the Q10 with the *CBR* mode gives the smallest degradation of recognition accuracy and an acceptable RTF.

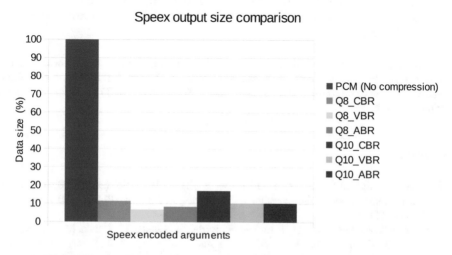

Fig. 4. Comparison of different settings for Speex speech data compression

5 Conclusion and Discussion

This paper aimed at summarizing the key research and development issues, and showed the recent performance of a Thai open-domain ASR system at NECTEC. Following the advanced algorithms on robust speech feature extraction, discriminative training, and multi-conditioned noisy speech training clearly raised the overall recognition accuracy on our real-noisy speech evaluation data. The novel hybrid word-subword language modeling method was shown to be highly efficient for making the system largely covers Thai lexical words at a small resource required. The system architecture was well designed to be ready for service deployment. The developed run-time system produced an acceptable response time and is scalable up on the user requirement.

Limitations of the system are of course existing. One major issue is the coverage of proper names always created every day. Although the system has been built with open-domain in mind, real applications are often domain and vocabulary specific. Enlarging the system lexicon is not always right but with the current highly covered lexicon, a method to rapidly adapt the system to cope with such specific set of vocabulary is more attractive. Another issue is the 4-gram rescoring part, which can clearly increase the overall recognition accuracy, has been skipped in our run-time system to preserve a low RTF. There might be a better way to take the larger n-gram into account. In real applications, the system robustness against a variety of background noise is still open for research. Background music and speaker separation is needed to make the system usable.

Deep neural network (DNN) has been in the recent trend of modern ASR as it naturally handles the large variation of input speech. The values of DNN hidden layers have also been proven to be an efficient features for further conventional processing. Recurrent neural network (RNN) and Long short-term memory (LSTM) has also been investigated for modern language modeling as their better properties to capture long dependency than the conventional n-gram model. Our current research focuses on such DNN-based algorithms. The RNN language model has also been experimented for our future ASR system [12].

Acknowledgements. We would like to thank Dr. Atiwong Suchato from Chulalongkorn University and his team for research collaboration under this project, and Dr. Chiori Hori from National Institute of Information and Communications Technology (NICT) Japan for her help in visiting research.

References

1. Saon, G., Kuo, H.J., Rennie, S., Picheny, M.: The IBM 2015 English conversational telephone speech recognition system. In: Proceedings of INTERSPEECH 2015, Dresden, Germany (2015)
2. Schalkwyk, J., Beeferman, D., Beaufays, F., Byrne, B., Chelba, C., Cohen, M., Garrett, M., Strope, B.: Google search by voice: a case study. In: Advances in Speech Recognition: Mobile Environments, Call Centers and Clinics, pp. 61–90. Springer, New York (2010)

3. Shaik, M., Tüske, Z., Tahir, M., Nussbaum-Thom, M., Schlüter, R., Ney, N.: Improvements in RWTH LVCSR evaluation systems for Polish, Portuguese, English, Urdu, and Arabic. In: INTERSPEECH 2015, Dresden, Germany, pp. 3154–3157 (2015)
4. Kasuriya, S., Sornlertlamvanich, V., Cotsomrong, P., Kanokphara, S., Thatphithakkul, N.: Thai speech corpus for speech recognition. In: Oriental COCOSDA 2003, Singapore (2003)
5. Saykham, K., Chotimongkol, A., Wutiwiwatchai, C.: Online temporal language model adaptation for a Thai broadcast news transcription system. In: LREC 2010, Valletta, Malta (2010)
6. Chotimongkol, A., Thatphithakkul, N., Purodakananda, S., Wutiwiwatchai, C., Chootrakool, P., Hansakunbuntheung, C., Suchato, A., Boonpramuk, P.: The development of a large Thai telephone speech corpus: LOTUS-Cell 2.0. In: Oriental COCOSDA 2010, Kathmandu, Nepal (2010)
7. Chotimongkol, A., Chunwijitra, V., Thatphithakkul, S., Kurpukdee, N., Wutiwiwatchai, C.: Elicit spoken-style data from social media through a style classifier. In: Oriental COCOSDA 2015, Shanghai, China (2015)
8. Chotimingkol, A., Saykham, K., Thatphithakkul, N., Wutiwiwatchai, C.: Toward benchmarking a general-domain Thai LVCSR system. In: ECTI-CON 2010, Thailand (2010)
9. Universal Speech Translation Advanced Research (U-STAR) consortium, http://www.ustar-consortium.com/
10. Wutiwiwatchai, C., Thangthai, K., Sertsi, P.: Thai ASR development for network-based speech translation. In: Oriental COCOSDA 2012, Macau, China (2012)
11. Thangthai, K., Chotimongkol, A., Wutiwiwatchai, C.: A hybrid language model for open-vocabulary Thai LVCSR. In: INTERSPEECH 2013, Lyon, France (2013)
12. Chunwijitra, V., Chotimongkol, A., Wutiwiwatchai, C.: Combining multiple-type input units using recurrent neural network for LVCSR language modeling. In: INTERSPEECH 2015, Dresden, Germany (2015)
13. Kurpukdee, N., Sertsi, P., Chunwijitra, S., Chunwijitra, V., Chotimongkol, A., Wutiwiwatchai, C.: Enhance run-time performance with a collaborative distributed speech recognition framework. In: ICSEC 2015, Thailand (2015)
14. Povey, D., Ghoshal, A., Boulianne, G., Burget, L., Glembek, O., Goel, N., Hannemann, M., Motlicek, P., Qian, Y., Schwarz, P., Silovsky, J., Stemmer, G., Vesely, K.: The Kaldi speech recognition toolkit. In: ASRU 2011, Hawaii, US (2011)
15. Stolcke, A.: SRILM—an extensible language modeling toolkit. In: ICSLP 2002, Colorado, US (2002)
16. El-Desoky, A., Gollan, C., Rybach, D., Schlüter, R., and Ney, H.: Investigating the use of morphological decomposition and diacritization for improving Arabic LVCSR. In: INTERSPEECH 2009, pp. 2679–2682. Brighton, UK (2009)
17. Kwon, O.W., Park, J.: Korean large vocabulary continuous speech recognition with morpheme-based recognition units. Speech Commun. **39**(3), 287–300 (2003)
18. Jongtaveesataporn, M., Thienlikit, I., Wutiwiwatchai, C., Furui, S.: Lexical units for Thai LVCSR. Speech Commun. **51**(4), 379–389 (2009)
19. Aroonmanakul, W.: Collocation and Thai word segmentation. In: SNLP-Oriental COCOSDA 2002, pp. 68–75. Prachuapkirikhan, Thailand (2002)
20. Haeb-Umbach, R., Ney, H.: Linear discriminant analysis for improved large vocabulary continuous speech recognition. ICASSP **1992**, 13–16 (1992)
21. Gopinath, R.: Maximum likelihood modeling with Gaussian distributions for classification. In ICASSP 1998, vol. 2, pp. 661– 664 (1998)
22. Bahl, L., Brown, P., de Souza, P., Mercer, R.: Maximum mutual information estimation of hidden Markov model parameters for speech recognition. In: ICASSP 1986, vol. 1, pp. 49–52 (1986)

23. Povey, D., Woodland, P.: Minimum phone error and i-smoothing for improved discriminative training. In: ICASSP, Kyoto, Japan (2012)
24. Speex: a free codec for free speech, http://www.speex.org/

Handwritten Character Strings on Medical Prescription Reading by Using Lexicon-Driven

Narumol Chumuang[1](✉) and Mahasak Ketcham[2]

[1] Department of Information Technology, Faculty of Information Technology,
King Mongkut's University of Technology North Bangkok, Bangkok, Thailand
lecho20@hotmail.com
[2] Department of Information Technology Management, Faculty of Information
Technology, King Mongkut's University of Technology North Bangkok,
Bangkok, Thailand
mahasak_k@it.kmutnb.ac.th

Abstract. In this paper, we present a handwritten character string recognition system on medical prescription reading. The medicine's words are recognized as an entire because there is no extra space between characters. The lexicon contains 520 medicine's words, which are stored in a trie structure. The recognition, the text line image is matched with the lexicon entries to obtain reliable segmentation and retrieve valid medicine's words. The pre-segmentation, the text line image is separated into aboriginal segments by connected component analysis. Lexicon matching, coherent segments are dynamically combined into candidate character patterns. The character classifier is embedded in lexicon matching to select characters matched with candidate pattern from dynamic series category. In the experiment result on 5200 handwritten character images achieved correct rate of 74.13%.

Keywords: Prescription reading · Handwritten character string recognition · Lexicon matching

1 Introduction

Medication error is event that can be prevented that's may be the cause or lead to inappropriate medication treat, harmful to the patient discrepancy which can made patients didn't receive the drug should have been [1, 2]. Example are prescribing a drug overdose, the physicians do not clearly specify the name drugs, prescribing does not match the patient's disease and so on. Medication error can be divided into four main issues such as (1) Prescription error, (2) Transcribing error, (3) Dispensing error and (4) Administration error [3]. The four main points have already said we can use technology to reduce deviations at any point. The research reveals that there are at present using technology to reduce medication errors like using the computerized provider order entry: CPOE used in hospitals to reduce medication errors in hospitals in the United States [4]. Bringing electronic devices to record medication order can be printed out directly [5]. Reducing medication errors to patients with natural language processing (NLP) [6, 7, 8]. The reduction medication errors by using image processing

[9]. This paper focus on apply the image processing technology combined with natural language processing to reduce transcribing error. The concept image reading process prescription drug list and then converted automatically. This makes the process of copying the prescription drug orders from paper forms into text in a database system.

The system for an automatic reading both of printed / handwritten has become a trending area of research in OCR (optical character recognition) applications [10, 11, 12, 13]. The printed one is easier than handwritten because it has exact pattern [14, 15]. The handwritten has various patterns that are variant the emotions. Moreover, the main reason for that is not only the challenge in simulating the human reading but also its utility in development document analysis like medical prescription form, postal address, live mail address, bank cheques, information sheets. Medical prescription is an order, especially by a physician, for the preparation and administration of a medicine, therapeutic regimen, assistive or corrective device, or other treatment in general health care and hospital [16]. In many developing automatic reading of handwritten, the present medical prescription processing procedure requires a pharmacist to read and manually enter the information present on medical prescription (or its image) and also verify the entries like Visit number (VN), patient's information, history of medicine allergy, medicine's information and physician's signature. They are a lot of things to consider and as a large number of medical prescriptions have to be process every day in a hospital. An automatic reading system can save timing and procedures of the work. However with the success accomplish in character recognition over the last few decades. The recognition of handwritten information and the verification of signatures present on document still remain a challenging problem in document image analysis [17, 18, 19].

The organization in this paper separate with section. In Sect. 2 describe about background knowledge. In Sect. 3 present our proposed algorithm with the diagram and describe thoroughly methodology. The experimental results and discuss conclusion are in Sects. 4 and 5 respectively.

2 Background Knowledge

2.1 Prescription

Prescription is the list of medicine treatments from physician to pharmacist. It consist both of printed and handwritten about the details of patient and physician, information of medicine and others. In this paper, we use OPD prescriptions form as show in Fig. 1 for our experiment.

3 Proposed Algorithm

The prescription reading process, the task of ROI and text line segmentation have been accomplished externally. Figure 2 gives the block diagram of our system. For simplicity of illustration, we assume the input to the system is a text line image.

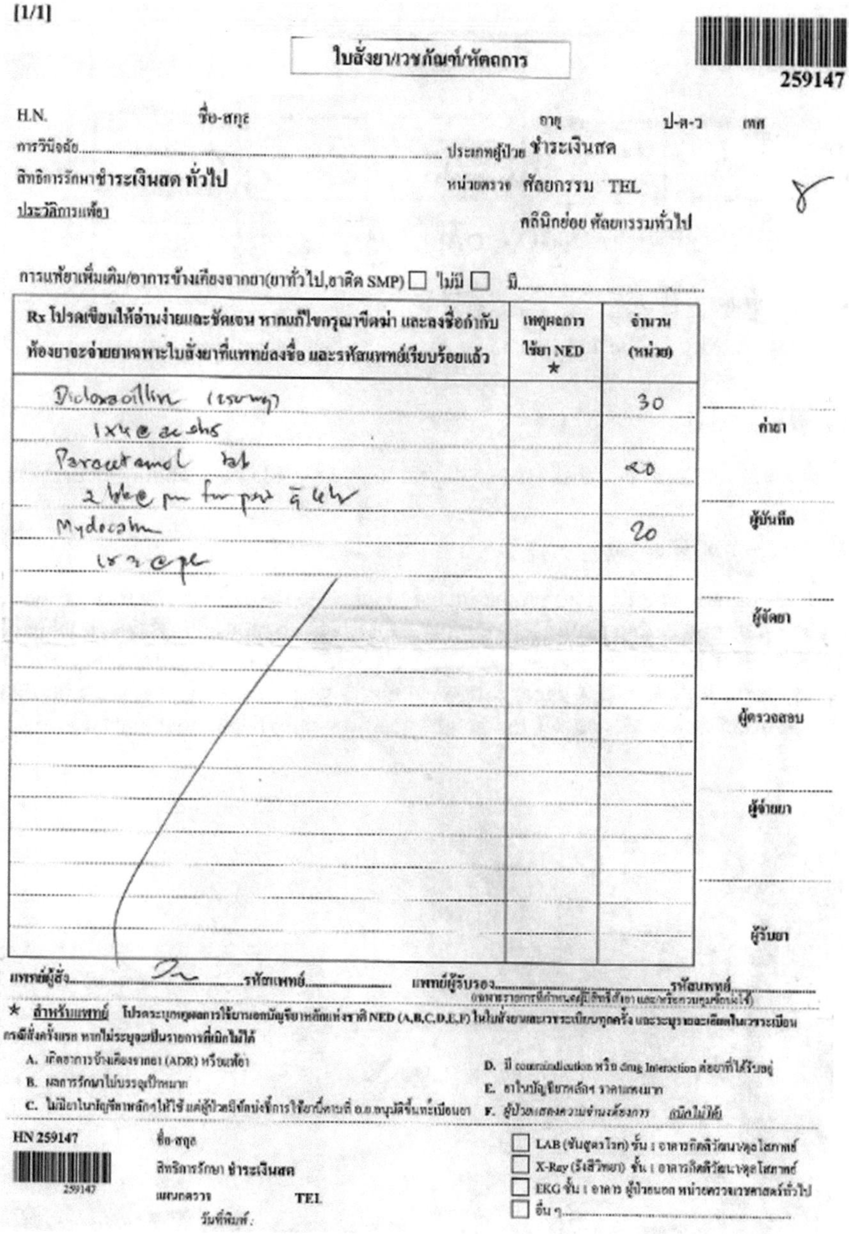
[1/1]

ใบสั่งยา/เวชภัณฑ์/หัตถการ

259147

H.N. ชื่อ-สกุล อายุ ป-ด-ว เพศ

การวินิจฉัย ประเภทผู้ป่วย ชำระเงินสด

สิทธิการรักษา ชำระเงินสด ทั่วไป หน่วยตรวจ ศัลยกรรม TEL

ประวัติการแพ้ยา คลินิกย่อย ศัลยกรรมทั่วไป

การแพ้ยาเพิ่มเติม/อาการข้างเคียงจากยา(ยาทั่วไป,ยาติด SMP) ☐ ไม่มี ☐ มี

Rx โปรดเขียนให้อ่านง่ายและชัดเจน หากแก้ไขกรุณาขีดฆ่า และลงชื่อกำกับ ห้องยาจะจ่ายยาเฉพาะใบสั่งยาที่แพทย์ลงชื่อ และรหัสแพทย์เรียบร้อยแล้ว	เหตุผลการใช้ยา NED ★	จำนวน (หน่วย)
Dicloxacillin (250 mg)		30
1x4 @ ac hs		
Paracetamol tab		20
2 tab @ prn for pain q 4 hr		
Mydocalm		20
1x3 @ pc		

ค่ายา ผู้บันทึก ผู้จัดยา ผู้ตรวจสอบ ผู้จ่ายยา ผู้รับยา

แพทย์ผู้สั่ง ... รหัสแพทย์ ... แพทย์ผู้รับรอง ... รหัสแพทย์ ...

★ สำหรับแพทย์ โปรดระบุเหตุผลการใช้ยานอกบัญชียาหลักแห่งชาติ NED (A,B,C,D,E,F) ในใบสั่งยาและเวชระเบียนทุกครั้ง

A. เกิดอาการข้างเคียงจากยา (ADR) หรือแพ้ยา
B. ผลการรักษาไม่บรรลุเป้าหมาย
C. ไม่มียาในบัญชียาหลักฯให้ใช้
D. มี contraindication หรือ drug interaction
E. ยาในบัญชียาหลักฯ ราคาแพงมาก
F. ผู้ป่วยแสดงความจำนงต้องการ (เบิกไม่ได้)

HN 259147 ชื่อ-สกุล สิทธิการรักษา ชำระเงินสด แผนกตรวจ TEL วันที่พิมพ์

☐ LAB ☐ X-Ray ☐ EKG ☐ อื่น ๆ

Fig. 1. Example of prescription

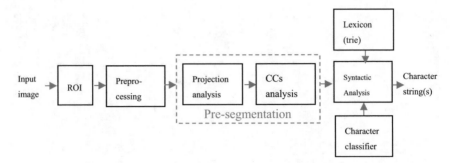

Fig. 2. Block diagram of a character string recognition system

3.1   Input Image

The prescription form is transferred to digital image.jpg file by scanning with 300 dpi.

3.2   Region of Interest

The scope of area is interested on the image to take the measure in the data or the ROI calculation or processing [20]. Determining ROI may use mouse, joystick, trackball, or light pen. ROI has many shapes like square, circle, oval or irregular type. The prescription form follow in this paper follow as Fig. 1 was set the interest are with fix the width position at 8–688 and set the height position at 267–604 that shows in Fig. 3.

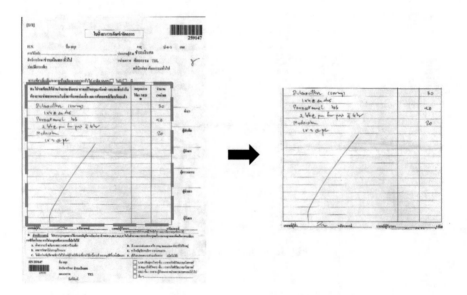

Fig. 3. Interest area is set on the prescription

3.3 Preprocessing

In this paper, we set standard size for prescription image to 1024 × 768 pixels and convert to gray scale with (1).

$$p = (0.3 \times R) + (0.59 \times G) + (0.11 \times B) \tag{1}$$

Set p is the prescription image. R, G and B represent red, green and blue, respectively. Then convert gray scale to binary with (2).

$$I(x, y) = \begin{cases} 1; p(x, y) \neq 0 \\ 0; p(x, y) = 0 \end{cases} \tag{2}$$

Assign $I(x, y)$ are the coordinates of pixel in the binary image and $p(x, y)$ represents the coordinates of pixel in the gray scale image, see in Fig. 4.

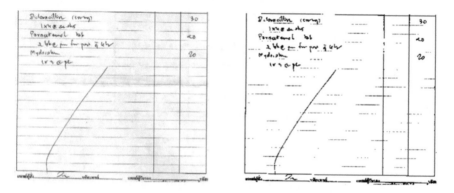

Fig. 4. Converting image to binary

3.4 Pre-segmentation

In pre-segmentation, the text line image is separated into characters or segments. Off-line handwritten character recognition is a difficult research and challenge problems in pattern recognition. However, incorrectly segmented characters will cause mis-classifications of characters which in turn may lead to wrong results. Therefore, two step for text segmentation were used to analysis.

- **Projection Analysis**

  The first important step is pre-segmentation. The use of the projection or histogram method simplifies the problem of character segmentation (Fig. 5).

- **Connected Components Analysis**

  The definition of adjacent pixels commonly are 8-connectivity (3) and in Fig. 6.

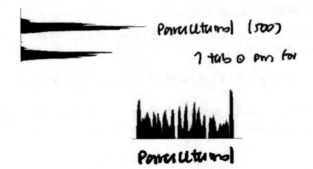

Fig. 5. Horizontal and vertical projection

$p_{(x-1,y-1)}$	$p_{(x,y-1)}$	$p_{(x+1,y-1)}$
$p_{(x-1,y)}$	$p_{(x,y)}$	$p_{(x+1,y)}$
$p_{(x-1,y+1)}$	$p_{(x,y+1)}$	$p_{(x+1,y+1)}$

Fig. 6. The 8-connectivity

$$N_8p = \left\{ \begin{array}{l} p_{(x+1,y)}, p_{(x-1,y)}, p_{(x,y+1)}, p_{(x,y-1)}, \\ p_{(x+1,y+1)}, p_{(x-1,y+1)}, p_{(x-1,y-1)} \end{array} \right\} \tag{3}$$

where N_8p is 8-connectivity and $p_{(x, y)}$ represents pixel's coordinates.

3.5   Character Classifier

This paper focus in automatic reading handwritten on prescription then we briefly our classification process. Multilayer perceptron algorithm used for character classification in this paper. The classification process, the high quality global features 1–9 [19] extraction were used. The matrix data of global features were input into perceptron in neuron network show in Fig. 7 for calculate output with (4).

In Fig. 8 illustrate the result character segmentation with connected components.

The character was sequential segmented. The number of 100 handwritten character images from A–Z and a–z were used to create character classes then the training set are 5200 images. In Fig. 8 first class is "P", second class is "a", third class is "r" etc.

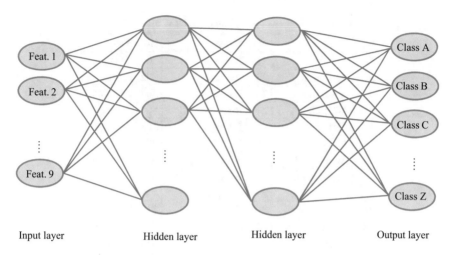

Input layer   Hidden layer   Hidden layer   Output layer

Fig. 7. The characters classified with MLP

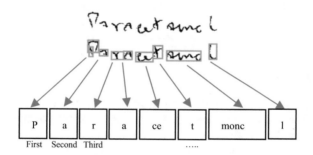

Fig. 8. Character segmentation with 8-connectivity

$$n = \sum_{i=1}^{z} x_i w_i + b \qquad (4)$$

where n is output from summation function, x_i is the matrix of global features i, w_i is weight of neuron i, z is the number of neuron in input layer, b is bias and i has value 1 to z. Fifty-two classes of A–Z and a–z were classified with accuracy is quite good.

3.6 Lexicon

International medicine names contains information about medications found in 185 countries around the world. In this paper the string length ranges from 5 to 28 and the average length is 17. The medicine's name with more than 20 characters have been trimmed to 20 characters. For creation database of the generic medication names. Index of A-Z total 520 popular medicine names such as aspirin, paracetamol, methadone etc. are show in Table 1.

Table 1. The example of medication names database

No.	Name	No.	Name		No.	Name
1	Abilify	21	Baclofen		501	Zanaflex
2	Acetaminophen	22	Bactrim		502	Zantac
3	Acyclovir	23	Bactroban		503	Zestoretic
4	Adderall	24	Belsomra	...	504	Zestril
5	Albuterol	25	Belviq		505	Zetia
6	Aleve	26	Benadryl		506	Ziac
7	Allopurinol	27	Benicar		507	Zithromax
8	Alprazolam	28	Biaxin		508	Zocor
9	Ambien	29	Bisoprolol		509	Zofran
10	Amiodarone	30	Boniva		510	Zoloft
11	Amitriptyline	31	Breo Ellipta		511	Zolpidem
12	Amlodipine	32	Brilinta		512	Zometa
13	Amoxicillin	33	Brintellix		513	Zostavax
14	Aricept	34	Bupropion		514	Zosyn
15	Aspirin	35	Buspar		515	Zovirax
...	....	...	...		...	...
20	Azithromycin	40	Bystolic		520	Zyvox

3.7   Syntactic Analysis

Determining syntax analysis about the positioning of the characters with probabilistic. The relationship of the first characters was to be taken into consideration with the next one follow as Fig. 9. We analysis syntactic the medicine's name with A–Z then our syntactic are 26 models from the popular 520 medicine's name.

4   The Experimental Results

The popular medicine's name total 520 dataset from A to Z alphabets are contained in database. In our experiment, we used five OPD prescriptions from a physician writing style. We found thirty-five medicine's name from our experiment for evaluating with accuracy present in Table 2. The average accuracy rate is 74.13%.

In Table 2 shown the accurate from prescription No. 1–5. The number of medicine's word show in "Total of words" row which are different number. Correct and error row are the correct and error rate respectively. In last row means the accuracy rate in each prescription. Two reasons made some the accurate are low because our pre-segmentation not clear. Another one is our lexicon used popular medicine's name then our system is known only 520 words only. Therefore, it is possible that prescription might have the medicine's name that are not in our lexicon.

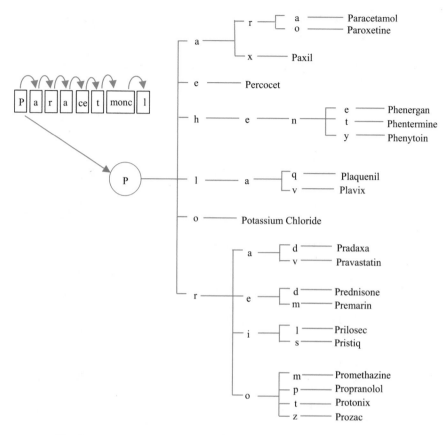

Fig. 9. A portion of the syntactic analysis with root P in lexicon (trie)

Table 2. The result of our algorithm arrange by the prescription

Prescription no.	1	2	3	4	5
Total of words	6	5	8	3	13
Correct	4	4	5	3	8
Error	2	1	8	0	5
Accuracy	66.66	80.00	62.50	100	61.53

5   Conclusion

The challenge of this work is handwritten which so difficult to make the computer understand its meaning. The contribution of this paper is present a handwritten character string recognition system on medical prescription reading maybe the first algorithm for automatic reading string text from handwritten medicine's name on prescription. The characters were separated in this step. Our character classification used MLP with reliable global feature for classified which it shown the accurate as satisfactory. The special models of medicine's lexicon were created for matching character with medicine's name word.

Acknowledgements. This research was funded by King Mongkut's University of Technology North Bangkok. Contract no. KMUTNB-60-GOV-29.

References

1. Haw, C.M., Geoff, D., Jean, S.: A review of medication administration errors reported in a large psychiatric hospital in the United Kingdom. In: Psychiatric Services, pp. 1610–1613 (2005)
2. Midlöv, P., Leila, B., Mehran, S., Peter, H., Eva, R., Tommy, E.: The effect of medication reconciliation in elderly patients at hospital discharge. Int. J. clin. pharm. **34**(1), 113–119 (2012)
3. Topping, J.: Errors of Observation and Their Treatment, vol. 62, pp. 72–144. Springer Science & Business Media, Berlin (2012)
4. David, C.R., Melanie, R.W., Lauren, E.W.O., Sarah, J.S., Mark, D.S., Bethany, B.: Reduction in medication errors in hospitals due to adoption of computerized provider order entry systems. J. Am. Med. Inform. Assoc. **20**(3), 470–476 (2013)
5. Benjamin, C.G., Robert, G., Kathryn, Y., Charles, A.: Reducing errors in discharge medication lists by using personal digital assistants. In: Psychiatric Services, pp. 1325–1326 (2014)
6. Fung, K.W., Chiang, S.J., Dina, D.-F.: Extracting drug indication information from structured product labels using natural language processing. J. Am. Med. Inform. Assoc. **20**(3), 482–488 (2013)
7. Jinsong, L., Cai, C.L., Joseph, G.C., Teng, C.C. Zhou, X. He, T. Harrison, D.J. Shah N. Sauer. B.C.: Performance of NLP tool designed to identify and extract biologic drug infusion data from clinical notes. Value in Health, **3**(17) (2014)
8. Doan, S., Bastarache, L., Klimkowski, S., Denny, J.C., Xu, H.: Integrating existing natural language processing tools for medication extraction from discharge summaries. J. Am. Med. Inform. Assoc. **17**(5), 528–531 (2010)
9. Caban, J.J., Rosebrock, A., Yoo, T.S.: Automatic identification of prescription drugs using shape distribution models. In: Image Processing (ICIP), 2012 19th IEEE International Conference on IEEE, pp. 1005–1008 (2012)
10. Afef, K., Asma, S., Abdel, B.: A system for an automatic reading of student information sheets. In: IEEE Computer Society International Conference on Document Analysis and Recognition, pp. 1265–1269 (2011)
11. Ramachandran, J., Satish, R.K., Prabhugouda, M.P., Umapada, P.: Automatic processing of handwritten bank cheque images: a survey. In: International Journal on Document Analysis and Recognition, vol. 15, pp. 267–296. Springer, Heidelberg (2012)
12. Vellingiriraj, E.K., Palanisamy, B.: Recognition of ancient tamil handwritten characters in palm manuscripts using genetic algorithm. Int. J. Sci. Eng. Technol. **2**(5), 342–346 (2013)
13. Amjad, R., Tanzila, S.: Neural networks for document image preprocessing: state of the art. In: Artificial Intelligent Revolution, pp. 253–273. Springer, Heidelberg (2014)
14. Satyaki, R., Ayan, C., Rituparna, P., Kaushik, G.: Printed text character analysis version-III: optical character recognition with noise reduction, background detection and user training mechanism for simple cursive fonts. Int. J. Inf. Eng. Electron. Bus. (IJIEEB) **7**(2), 27–37 (2015)

15. Lei, S., Qiang, H., Wei, J. Kai, C.: A robust approach for text detection from natural scene images. In: Pattern Recognition, pp. 2906–2920 (2015)
16. Babaei, N., et al.: Evaluation of the prescription skills of the fifth year dentistry program. Acad. J. Oral Dent. Med. **2**(2), 29–35 (2015)
17. Narumol, C., Preecha, K., Makasak, K.: The intelligence algorithm for character recognition on palm leaf manuscript. Far East J. Math. Sci. (FJMS) **89**(3), 333–345 (2015)
18. Singh, S., Kariveda, T., Gupta, J.D., Bhattacharya, K.: Handwritten words recognition for legal amounts of bank cheques in English script. In: Advances in Pattern Recognition (ICAPR), pp. 1–5 (2015)
19. Naruemol, C., Makasak, K.: Intelligent handwriting Thai signature recognition system based on artificial neuron network. In: IEEE TENCON Region 10 Conference, pp. 1–6. IEEE (2014)
20. Snook, L., Chris, P., Christian, B.: Voxel based versus region of interest analysis in diffusion tensor imaging of neurodevelopment. In: Neuron Image, pp. 243–252 (2007)

Part II

SSAI: Special Session on Artificial Intelligence

Eye Region Detection in Fatigue Monitoring for the Military Using AdaBoost Algorithm

Worawut Yimyam[1] and Mahasak Ketcham[2(✉)]

[1] Department of Computer Business, Phetchaburi Rajabhat University,
Phetchaburi, Thailand
worawut_yimyam@hotmail.com
[2] Department of Information Technology Management, King Mongkut's
University of Technology North Bangkok, Bangkok, Thailand
mahasak.k@it.kmutnb.ac.th

Abstract. This paper proposed the eye region detection in fatigue monitoring system based embedded system including smart phone devices and tablet for the use of military application. The eye movement and eye position tracking are detected for finding fatigue conditions. Haar-like feature and region of interest (ROI) are used for the fatigue analysis. From the experimental results, the algorithm works well in face and eye detection. The accuracy of the eye detection is 93%.

Keywords: Fatigue · Eye detection · Eye tracking

1 Introduction

Nowadays, in the execution of the military mission, there are important duties in several aspects such as the maintenance of independence, sovereignty and national security, the security surveillance along the border for monitoring the suspect and disaster. The prevention of the suppression of unlawful acts is also involved, such as drug trafficking, alienate worker, illegal deforestation, and forest ground invasion. Each mission requires a lot of force and takes a long time to carry out the mission, causing fatigue of military tasks. This fatigue is the causes of the accident while the military are on a mission. It reduces their driving ability such accident in the US military called M985 [1–3]. The investigation provided that the accident is due to the fatigue of the driver because he/she has to be worked for 73 cars. Moreover, it occurs in night when human eyes have the ability to perceive image only 6%. Another accident was a truck fall into a deep river, causing the driver died immediately. The incident was caused by the lack of sleep that affected the performance of decision, communication, and risk assessment, in which all are connected as a system [2–5]. The mission of the military is at risk all the time. It has worked for several hours in a task that required many military personnel. However, a number of military personnel in some agencies have increased. The soldiers and officers have to work many hours, thus the tiredness is occurred. Sometimes fatigue caused by the requirement of the agency itself. For example,

© Springer International Publishing AG 2018
T. Theeramunkong et al. (eds.), *Advances in Natural Language Processing,
Intelligent Informatics and Smart Technology*, Advances in Intelligent Systems

the military have to train as a pilot for flying helicopter. According to FFA rules [1–5], they have to sleep at least 9 h for working 24 h a day. The pilot needs to examine the helicopter before flying at 4 pm. If the pilot receives an order for the urgent mission, they will take off at 5:30 pm., land at 10:30 pm to the base, and then completing the mission in 00:30 am. The pilot may has less than an hour to sleep for preparing the flight path in a second day. After finishing the mission, many military may lose consciousness because of their hard working in 62 consecutive hours.

Currently, the technology developed for managing human fatigue and drowsiness in terms of the Intelligent Transportation System (ITS). It monitors the fatigue of driving that causes an accident on the highway by applying science technology to measure fatigue conditions. However, it still cannot detect the fatigue accurately [6, 7]. There are various methods for determining fatigue from the function of the eyes and twinkling. For example, Fazli et al. [8] proposed eye tracking for examining fatigue. Image processing performed the extraction of the eye position by the use of color for indentifying the eyelid whether it closed or opened. However, the closed and opened eyes depended on its characteristic in each person. The results contained some errors. Sun et al. [9] studied the fatigue monitoring system using sensors based physiological signals which combined eye and heart signal together. The results of the data transmission had problems in distance limitations. Edward et al. [10] analyzed the fatigue detection techniques and found that the technique involved with the examination of the eyes provided high reliability to determine the fatigue of the body. From the researches as mentioned above, there are some problems with the accuracy, reliability, and costs. The methods for the cost reduction such as Raudonis et al. [11] developed a device by connecting the glasses with the camera for reducing the distance of the eye position. It performed artificial neural network and matlab for classifying the characteristic of the eyes and the eye tracking. This worked well in classification; however, the system response was slow. The light conditions also affect to data analysis as explained in Chen and Kubo [12] who proposed face and rapid eye movement detection via webcam by applying Gabor filter. Gabor integrated filter with color for face detection. It provided high accuracy results. SQI method is proposed in Sung-Uk Jung [6] for removing irrelevant distractions to increase the performance of the eye detection. The conversion of the image to three-dimensional shape can help in finding the precise coordinates of the eye. In addition, the analyzing data technique, AdaBoost, can identify the position of the eye more efficiently. According to these researches, they are focus only on the monitoring of the function of the eyes without analyzing the fatigue that can causes accidents in real time. When the fatigue occurs, the system cannot predict what will happen further. Therefore, researchers developed the algorithm for monitoring the fatigue of the warriors based embedded system such as smart phone and tablet. Eye position detection and tracking are implemented in detecting the fatigue syndrome by Haar-like feature and region of interest. This paper is divided into five sections; section two provides preliminary, Sect. 3 proposes the algorithm for the fatigue monitoring system, Sect. 4 shows experimental results, and finally in section five is conclusion.

2 Preliminary

2.1 Image Processing Techniques

The digital image processing was employed in this research. Digital image processing is a signal processing method to transform analog signal to digital signal and shows in computer vision. The process also helps decreasing the noises of the image. The image from the processing will be described in spatial coordinates on two-dimension plane. The x and y axis describe the width and length of the image. Any points on (x, y) is called pixel.

2.2 Review Literature and Related Research

Akom [13] developed an algorithm to detect eyes. The detection focused on the partition of eye and pupil area. Once the system detected the eye, it will define the darker area as the pupil. However, it is difficult to measure the exact size of an eye or a pupil because of several factors such as the distance between eyes and camera, eye size, light intensity, etc. Discrete Wavelet Transform algorithm was employed in the experiment to find the exact position of the eye on the image. The algorithm works well in day time without sunglasses.

Hoang Le et al. [14] proposed eye blinking detection method with the smart glasses. To classify blinking, closed, or opened eye, the Gradiant Boosting method was used and compared the images from the smart glasses. However, there were still problems of this technique, for instance, if the image was dark, the system cannot detect and classify the eye movement. Therefore, the researchers add another method to detect the eye position more accurately, which is called Haar Cascade from OpenCV.

YenWei Chen and Kenji Kubo [12] investigated and developed face detection system and rapid eye movement by using Gabor filter method and web camera. This method combines filters with the color information to extract the face area in terms of geometric form. The method will detect the movement continuously and shows the result of the image on the monitor. The method showed the robust eye movement detection.

3 Proposed Method

3.1 Input Data

In first step, an input data is obtained from a smart phone for implementing in next procedure.

3.2 Image Pre-processing

Image enhancement is important for pre-processing process. The RGB color image is converted to Gray-scale image. The calculation is expressed below.

$$\text{Gray} = 0.299 \times R + 0.587 \times G + 0.114 \times B \tag{1}$$

where

G = Gray intensity value is between 0 and 255
R = Red intensity value is between 0 and 255
G = Green intensity value is between 0 and 255
B = Blue intensity value is between 0 and 255

3.3 Eye Region Detection

Haar-like feature proposed by Viola and Jones [15]. Haar-like feature extraction is a popular technique implemented for finding the difference of the intensity between white and black areas (Fig. 1).

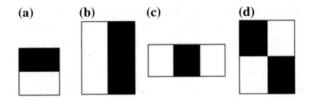

Fig. 1. **a** Haar-like feature characteristics, **b** Two-rectangle features. **c** Three-rectangle features. **d** Four-rectangle features

Integral image is employed to calculate the area for searching Haar-like feature. Firstly, summing the vertical and horizontal intensity can derive integral image. The intensity in D area is calculated by using one, two, three, and four as the reference points. The one equals to the sum of the intensity of the A area. The two equals to the sum of the intensity of the A and B areas. The three is equal to the sum of the intensity of the A and C areas. The four is equal to the sum of the intensity of the A, B, C, and D areas. Thus, the sum within D area can be calculated as Eq. (2).

$$D = 4 - 3 - 2 + 1. \tag{2}$$

From Fig. 2, the value of the sum of the intensity within D area equals to five that is utilized to find Haar-like feature. Six reference points are applied for two adjacent squares. The reference points including eight and nine is employed for three and four adjacent squares. Every square size has the same computation time, thus a strong point of an integral image affects the computation time.

AdaBoost Algorithm. Haar-like feature provides a large number of features. However, a small number of the features are more preferred [15]. Thus, AdaBoost algorithm is integrated for improving features. It finds weak classifiers and combines them as a strong classifier.

- Let images are $(x_1, y_1), \ldots, (x_n, y_n)$, where $y_i = 0, 1$ for negative and positive image, respectively.

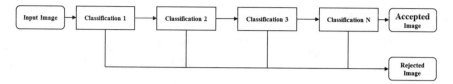

Fig. 2. Cascade classification

- Define the first weight as the Eq. (2).

$$W_{1,i} = \frac{1}{2m}, \frac{1}{2l},\tag{3}$$

  where $y_i = 0, 1$,
  m is the number of negative image.
  l is the number of positive image.

- Let $t = 1, \ldots, T$

  (a) Normalize weights

$$w_{t,i} \leftarrow \frac{w_{t,i}}{\sum_{j=1}^{n} w_{t,j}},\tag{4}$$

  where w_t is a probability distribution.
  (b) Compute a weak classification function (h_j) and an error $(\in_j)$ in each feature (j)

$$h_j(x_i) = \begin{cases} 1, & \text{for } p_j f_j(x) < p_j \,\theta_j, \\ 0, & \text{otherwise}, \end{cases}\tag{5}$$

$$\in_j = \sum_i w_i |h_j(x_i) - y_i|.\tag{6}$$

  (c) Select the weak classification with the lowest error
  (d) Adjust weights

$$w_{t+1,i} = w_{t,i} \beta_t^{1-e_i},\tag{7}$$

  where
  $e_i = 0$, When x_i is classified correctly
  $e_i = 1$, Otherwise
  And

$$\beta_t = \frac{\in_t}{1 - \in_t}.\tag{8}$$

- Calculate the final classification as

$$h(x) = \begin{cases} 1, & if \ \sum_{t=1}^{T} \alpha_t h_t(x) \geq \frac{1}{2}\sum_{t=1}^{T} \alpha_t, \\ 0, & \text{otherwise}, \end{cases} \quad (9)$$

where

$$\alpha_t = \log\frac{1}{\beta_t}.$$

Cascade classification. After receiving the strong classifier equation, cascade classification divides it into n stages, which mean that the inputs have not required computing in all stages. However, the inputs that pass through all stages will be recognized. Otherwise, it will be rejected.

Eye Detection is shown in Fig. 3.

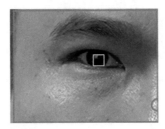

Fig. 3. The eye position is detected by the use of Cascade Classifier from Open CV

3.4 Pupil Detection

Pupil detection applies region of interest (ROI) to create a circle around pupil position. The circle function of Open CV is utilized for receiving the center point of the pupils (x, y) and then determining the radius (r) which is referred to the circle equation as shown in Eq. (10) (Fig. 4).

$$\text{radius } (r) = \sqrt{(x - Center\_x)^2 + (y - Center\_y)^2} \quad (10)$$

Fig. 4. The pupil position is detected by the use of ROI technique

3.5   Eye Tracking

Eye tracking based template matching is the process for tracking the directions of the eye movement in order to find the eye fatigue conditions such as twinkling, and the distance of the eyelid. An initial coordinates used for the movement can be calculated as the equation below;

$$(x_c, y_c) = \left(\frac{x_r + x_l}{2}\right), \left(\frac{y_r + y_l}{2}\right) \tag{11}$$

where
$(x_c, y_c) =$ Center coordinate between the eyes
$(x_r, y_r) =$ Right eye coordinate
$(x_r, y_r) =$ Right eye coordinate

After obtaining initial coordinates, the change of the position of any point to other positions affects the distance and movement of the cursor. The distance is calculated in Eq. (12).

$$distance = \sqrt{\left(x_c - x_c'\right)^2 + \left(y_c - y_c'\right)^2} \tag{12}$$

where

$$x_c, y_c = \text{center coordinate between the pupils}$$

New coordinate from the mouse movement can be calculated from the Eq. (13), (14), (15), and (16).

$$\theta = \arctan\left(\frac{y_c - y_c'}{x_c - x_c'}\right) \tag{13}$$

$$x_{c\_update} = x_c + \frac{distance}{3} \cos\theta \tag{14}$$

$$y_{c\_update} = y_c + \frac{distance}{3} \sin\theta \tag{15}$$

$$(x_{c\_update}, y_{c\_update}) = \text{coordinate of mouse} \tag{16}$$

Match template function of OpenCV is operated with Motion Event on android mobile (Fig. 5).

3.6   Face Detection

Face detection examines a region of face in the received image by using Cascade Classifier library of OpenCV for detection. The face detection considers the detection of the eyes by comparing Histogram values of the left eye and the right eye within the

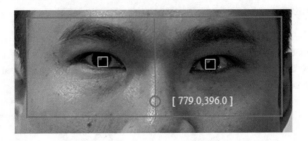

Fig. 5. An example of initial coordinates referred to the movement

image. If the system can detect both two eyes, it will create a square box around the face region.

4 Experimental Results

The experiment was examined via smart phone for 100 times. The details are described below.

Eye Detection Testing: Eye detection testing processes the image receiving and location tracking procedures while starting the program

Eye Movement Testing: Eye movement testing verifies the directions of the eye movement including,

- The system tracks at the left side when the eyes turn left.
- The system tracks at the right side when the eyes turn right.
- The system tracks at the topside when the eyes move to above.
- The system tracks at the bottom side when the eyes move to below.

Face Detection Testing

Face detection testing performs the system when users move their head. Its performance testing for analyzing the accuracy is shown in Eq. (17).

$$A = \frac{Nc}{N} \times 100 \qquad (17)$$

A   is percentage of accuracy.
Nc   is number of accurate processing.
N   is number of all the experiments

The experiment is divided into two types; performance testing for the eye detection and eye tracking (cursor control). The results are shown as below (Tables 1 and 2).

From the experiments that have tested for 100 times, we found the accuracy of the eye detection and eye tracking performances are described as follow (Fig. 6).

The accuracies from the eye detection capability when the eyes view straight on, look at the top, look at the bottom, turn to the left, and turn to the right are 96, 87, 89, 96, 95, and 93 percents respectively. The accuracies from the eye tracking capability when the eyes look at the top, look at the bottom, turn to the left, turn to the right, and view straight on are 73,76,69, 98, and 78 percents respectively (Fig. 7).

Table 1. Experimental results of the performance of the eye detection

| | The position of the eye detection | | | | | Mean |
	Straight	Top	Bottom	Left	Right	
Accuracy	0.96	0.87	0.89	0.96	0.95	0.93

Table 2. Experimental results of the performance of the eye tracking

| | The position of the eye tracking | | | | | Mean |
	Top	Bottom	Left	Right	Straight	
Accuracy	0.73	0.76	0.69	0.75	0.98	0.78

Fig. 6. **a** The experiment of the face detection and eye tracking at the straight view. **b** The experiment of the face detection and eye tracking at the left view. **c** The experiment of the face detection and eye tracking at the right view. **d** The experiment of the face detection and eye tracking at the top view. **e** The experiment of the face detection and eye tracking at the bottom view

Fig. 7. Results of the eye detection

5   Conclusion

This paper proposed the eye region detection in fatigue monitoring for the Military using AdaBoost Algorithm. The eye position and eye movement are the main factors leading to the eye detection and pupil tracking for finding the face and analyzing the fatigue conditions in further procedure. The Haar-like feature and region of interest (ROI) are used for the fatigue detection. From the experimental results, the algorithm works well in face and eye detection. The overall accuracy of eye detection is 93% and the accuracy of eye tracking is 78%. The improvement of the algorithm's robustness will be focused in further research in order to handle with lighting and the distance of the camera that result to processing performance in fatigue analysis.

References

1. Sicard, B.: Risk propensity assessment in military special operations. Mil. Med. **166**(10), 871 (2001)
2. Nakagawa, T., Kawachi, T., Arimitsu, S., Kanno, M., Sasaki, K., Hosaka, H.: Drowsiness detection using spectrum analysis of eye movement and effective stimuli to keep driver awake. In: DENSO Tech. Rev. **12**(1) (2006)
3. US Army Safety Center: Sustaining Performance in Combat, Flight fax, (31)5:9–11(2003)
4. Lin, C.T., Ko, L.W., Chung, I.F., Huang, T.Y., Chen, Y.C., Jung, T.P., Liang, S.F.: Adaptive EEG-based alertness estimation system by using ICA-based fuzzy neural networks. In: IEEE transactions on circuits and systems. **53**(11) (2006)
5. Cai, H, Lin, Y.: An experiment to non-intrusively collect physiological parameters towards driver state detection. In: Proceedings of SAE 2007 World Congress, No.2007-01-0403. SAE Technical Paper, (2007)
6. Jung, S. U., Yoo, J.H.: Robust eye detection using self quotient image. In: Intelligent Signal Processing and Communications. ISPACS'06, pp. 263–266. IEEE, Japan (2006)
7. Brandt, T., Stemmer, R., Rakotonirainy, A.: Affordable visual driver monitoring system for fatigue and monotony. In: Systems, Man and Cybernetics, IEEE International Conference on, Vol. 7, pp. 6451–6456. IEEE, (2004)
8. Fazli, S., Esfehani, P.: Tracking eye state for fatigue detection. In International Conference on Advances in Computer and Electrical Engineering In: ICACEE 2012, pp. 17–20. (2012)
9. Sun, Y., Yu, X., Berilla, J.: An innovative non-invasive ecg sensor and comparison study with clinic system. In: Bioengineering Conference (NEBEC), 39th Annual Northeast, pp. 163–164. IEEE. (2013)
10. Edwards, D.J., Sirois, B., Dawson, T., Aguirre, A., et al.: Evaluation of fatigue management technologies using weighted feature matrix method. In: Processing of the Fourth International Driving Symposium on Human Factors in Driver Assessment, Training and Vehicle Desigr, Stevenson, Washington(2007).
11. Raudonis, V., Simutis, R., Narvydas, G.: Discrete eye tracking for medical applications. In: Applied Sciences in Biomedical and Communication Technologies, ISABEL 2009. 2nd International Symposium on. pp. 1–6. IEEE. (2009)
12. Chen, Y.W., Kubo, K.: A robust eye detection and tracking technique using gabor filters. In: Intelligent Information Hiding and Multimedia Signal Processing, 2007. IIHMSP 2007. Third International Conference on vol. 1, pp. 109–112. IEEE (2007)

13. Akom, M.: Face and eye tracking for drowsiness detection using image processing method. In: Department of Electeical Engineering Faculty of Engineering Kasersart University Bangkok. (2552)
14. Hoang, L., Thanh, D, Feng, L.: Eye blink detection for OKsmart glasses. In: IEEE International Symposium on Multimedia. pp. 305–308. IEEE (2013)
15. Viola, P., Jones, M.: Rapid object detection using a boosted cascade of simple features. In Computer Vision and Pattern Recognition, 2001. CVPR 2001. Proceedings of the 2001 IEEE Computer Society Conference on (Vol. 1, pp. I-I). IEEE

The Feature Extraction of ECG Signal in Myocardial Infarction Patients

Mahasak Ketcham and Anchana Muankid[✉]

Faculty of Information Technology, King Mongkut's University
of Technology North Bangkok, Bangkok, Thailand
mahasak.k@it.kmutnb.ac.th, muan.anchana@gmail.com,
s5607011910072@email.kmutnb.ac.th

Abstract. Cardiovascular disease is one of the most serious diseases in the world. The physicians use electrocardiogram (ECG) to detect and diagnosis cardiovascular diseases. To enhance the ECG analysis performance, quality of the ECG signal needs to be improved. This paper proposes an ECG analysis algorithm using Wavelet transform to classify Myocardial Infarction patients. Steps in the ECG signal analysis are noise elimination of ECG signal, R peak Detection, QRS Complex Detection and Myocardial Infarction Classification. The results showed that the accuracy of classification equaled 75%, the sensitivity was 80% and the specificity was at 77.78%.

Keywords: ECG analysis · Wavelets transform · Myocardial infarction

1 Introduction

Cardiovascular disease (CVD) is the most common cause of death in the world [1]. There are many types of Electrocardiogram (ECG) research in cardiovascular disease patients. The research was divided into two main types by disease, Arrhythmia and Myocardial Infarction (MI). Arrhythmia is the disease that has heart rhythm problems, for example, Bradycardia, Tachycadia and Atrial Flutter. This research group focuses on electrocardiogram signal analysis to extract the rhythm of heart and classify patients from healthy controls. Another research group is Myocardial Infarction (MI). A Myocardial Infarction (MI) is a cardiovascular disease that has insufficient blood flow to the heart muscle, the heart muscle cells are death and might cause acute heart failure, and cardiac arrest. Cause of Myocardial Infarction is shown in Fig. 1.

The process of Myocardial Infarction (MI) classification is divided into three stages, preprocessing stage; eliminate electrocardiogram signals noise before the analysis, waveform detection stage; detect R peak, QRS complex and ST segment and Myocardial Infarction classification stage; extract the feature of detected waveform and classify Myocardial Infarction patients from health control.

The most complex process is Myocardial Infarction classification stage as varies of electrocardiogram waveform pattern [3] and electrocardiogram varies in time [4]. The Fig. 2 show examples of Myocardial Infarction waveform patterns, A is normal,

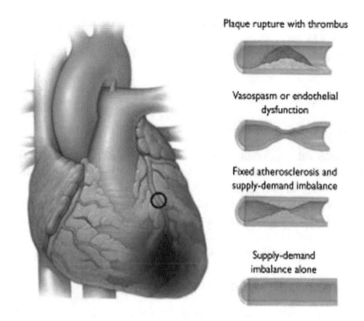

Fig. 1. Cause of myocardial infarction [2]

B is early hyper-acute T wave, C is hyper-acute T wave, D is ST-segment elevation, E is Giant R Peak and F is ST-segment elevation [3].

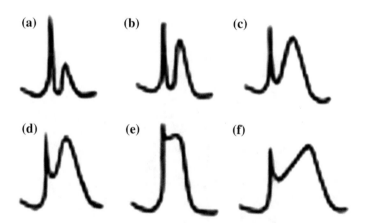

Fig. 2. Myocardial infarction waveform patterns [3]

This paper purposes an electrocardiogram feature extraction algorithm using Wavelet transform to classify Myocardial Infarction patients from healthy controls. This is the basic algorithm for identify cell death position. The results used to diagnostic decision support of physician. The three steps of electrocardiogram signal analysis are noise elimination, feature extraction and Myocardial Infarction classification. Accuracy, Sensitivity and Specificity are used to evaluate the algorithm.

The remaining of the paper is organized as follows: the first section is Introduction, Sect. 2 covers Electrophysiology, Wavelet Transform and related works, Sect. 3 is Electrocardiogram Analysis, Sect. 4 is Results and Sect. 5 is the Conclusion.

2 Background and Notations

2.1 Electrical Conduction System

The initial electrical conduction system is the Sinoatrial (SA) node. The electrical impulse from SA node spreads through the Internodal Pathway to the Atrioventricular (AV) node, passing the bundle branches down to the Purkinje fiber [5].

The depolarization is a process of electrical impulses propagating through the cardiac tissue, contracting the heart muscles. The re-polarization starts in the opposite direction. The electrical activity of the heart is represented by the electrocardiogram (ECG) which is detected by skin electrodes.

2.2 Electrocardiogram Signal

Electrocardiogram (ECG) is a graph showing the heart electrical activity [6] that's detected by attaching electrodes to the skin. An ECG normal waveform, the P wave occurs first, followed by the QRS complexes and the T wave. The ranges between the waves are called segments. The X-axis shows the record speed (millimeters/second), and the Y-axis shows the energy (amplitude). A normal ECG waveform is shown in Fig. 3.

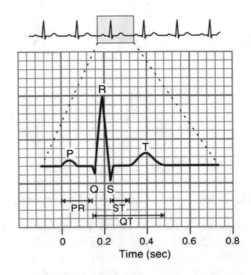

Fig. 3. Normal ECG waveform [7]

The ECG signal is detected by attaching electrodes to the skin. The ECG lead placements are divided into three types, Limb Leads (Bipolar), Augmented Limb Leads (Unipolar) and Precordial Leads (Unipolar) [8]. The Limb Leads and the Augmented Limb Leads are the view of signal from electrodes that are attached to the limb, consisting of Lead I, II, III, aVL, aVR and aVF. There are six views obtained from limb leads. The Precordial Leads are the view of signal form six chest electrodes, consisting of V1, V2, ..., V6. The various characteristic features of ECG are used to classify the cardiac abnormalities and support decisions by physicians.

2.3   Wavelet Transform

A wavelet based signal technique is an effective tool for nonstrationary ECG signal analysis and characterization of local wave (P, T and QRS complex morphologies) [7]. Even if a signal is not represented well by one member of the Daubechies family, it may still be efficiently represented by another [9]. This paper, selection of the wavelet decomposition of ECG signal at level 4, using Daubechies4 is then undertaken since this waveform resembles the original ECG signal.

2.4   Related Works

ECG varies in time, researchers have developed a computerize system to monitor patients heart health accurately and easily [4]. The ECG analysis topic research is divided into three main areas, preprocessing stage, waveform detection and cardio-vascular disease (CVD) classification. Details are as follows;

Preprocessing stage. The researches in this group concentrated on the preparation of ECG signals before the analysis. The researchers focus on noise elimination; for example, eliminate power-line interference of ECG signal by Radial basis function (RBF) wiener hybrid filter [10].

Waveform Detection. The researches in this group concentrated on the detection of R peak, QRS complex and ST segment. The significant waveform is then used to classify CVD and identify the heart abnormality location. The most popular technique is Pan-Tompkins Algorithm [11] which is used to detect R peak and identify QRS complex. First step is an integer coefficient bandpass filter. Next step is derivative approximate then, applying squaring function and Moving-Window integration. After fiducial mark on ECG signal, adjusting the thresholds, average RR Interval and rate limits. The last step is T-wave Identification.

Cardiovascular Disease Classification. The researches in this group concentrated on classifying health control and Cardiovascular Disease patients. This research group was divided into two types, Arrhythmia and Myocardial Infarction (MI). The most common technique used for classification is data mining such as Neural Network, Least Square Support Vector Machine (LS-SVM) [4, 12] and Multi-layer back propagation neural network [13].

The survey of the ECG signal used in the analysis of Myocardial Infarction shows that the classification of Myocardial Infarction by ECG waveform analysis based on Support vector machines (SVMs) and Gaussian mixture model (GMMs) provides 82.50% accuracy [14]. Banerjee analyzed the ECG signal from PTB-DB and classified

IM patients using a Cross Wavelet Transform (XWT) technique [15] which provided accuracy up to 92.50%. The PTB-DB contains an ECG data associated with Myocardial Infarction and 15 leads of ECG signal, PTB-DB is appropriate for this research purpose [16]. The studies are appended by adjusting the XWT parameters for higher accuracy. The classification accuracy is up to 97.6% [17].

3   Proposed Method

The steps in ECG signal analysis are noise elimination from the ECG signal, R peak Detection, QRS Complex Detection and Myocardial Infarction Classification. The ECG analysis process is shown in Fig. 4.

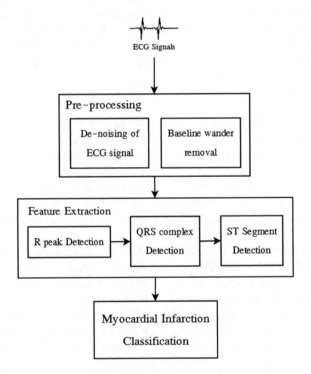

Fig. 4. ECG analysis process

3.1   Preprocessing Stage

ECG signal are usually contaminated by various types of noise, Power Line Interference, Electrode Contact Noise, Motion Artifact, Muscle Contraction and Base Line Wander. The ECG signal data from the PTB Diagnostic ECG database is used to analyze ECG signal waveform [18]. The samples are 10 healthy controls and 10 Myocardial Infarction patients, using 10 s of ECG signal in Lead I. The original ECG signal is shown in Fig. 5.

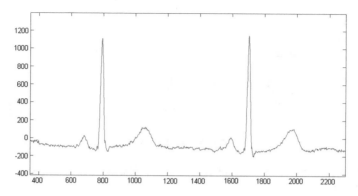

Fig. 5. Original ECG signal

At this stage, the noise is removed to improve ECG signal quality [19]. A wavelet based signal technique is an effective tool for nonstrationary ECG signal analysis and characterization of local wave (P, T and QRS complex morphologies) [7]. A multilevel one-dimensional wavelet analysis using a Daubechies family in MATLAB is then conducted. Selection of the wavelet decomposition of ECG signal at level 4, using Daubechies4 is then undertaken since this waveform resembles the original ECG signal. Daubechies series of ECG signals are shown in Fig. 6.

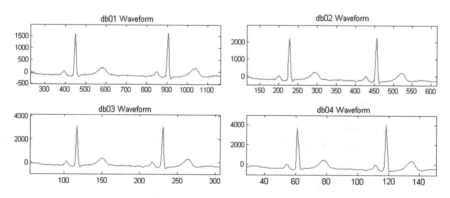

Fig. 6. Daubechies4 waveform

3.2 R-Peak Detection

R-peak detection is the most important task in ECG signal analysis as there is an obvious peak detected first [20]. This step uses a threshold based algorithm to detect the R-peak. Assigned threshold is 2000 since every R-peak of Daubechies4 ECG signal is greater than 2000. The peak that value more than 2000 is R-peak. The R-peak detection equations are as follows.

Defined:

Ra   is R peak amplitude,

Rx   is R peak position

$$Rxp = Rx - 1 \tag{1}$$

$$Rxn = Rx + 1 \tag{2}$$

$$R\,peak = (Rx, Ra) \tag{3}$$

where $Rx > (1)$ and $Rx > (2)$ and $Ra \geq 2000$. The results of R-peck detection is shown in Fig. 7.

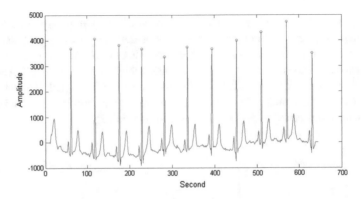

Fig. 7. R-peak detection result

3.3 QRS Complex Detection

After R-peak detection, QRS complex identifies by calculating the lowest point in the previous position and the next position. The QRS complex detection equations are as follows.

Defined:

Qa   is Q peak amplitude,

Qx   is Q peak position

$$Qxp = Qx - 1 \tag{4}$$

$$Qxn = Qx + 1 \tag{5}$$

$$Q\,peak = (Qx, Qa) \tag{6}$$

where $Qx < (4)$ and $Qx < (5)$ and $Qx < Rx$.

Defined:

Sa   is S peak amplitude,

Sx   is S peak position

$$Sxp = Sx - 1 \tag{7}$$

$$Sxn = Sx + 1 \tag{8}$$

$$S \text{ peak} = (Sx, Sa) \tag{9}$$

where $Sx < (7)$ and $Sx < (8)$ and $Sx < Rx$

The detection of QRS complex is shown in Fig. 8.

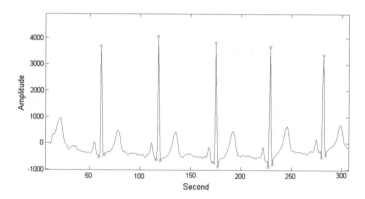

Fig. 8. QRS complex identification result

3.4   ST Segment Detection

Choosing the appropriate features are very important because this will affect the classification of the ECG signal. If the feature is not appropriate, classification results in an error [21]. This paper covers invert T-wave feature when classifying Myocardial Infarction patients. The S-T segment detection equations are as follows.

Defined:

Ta   is T wave amplitude,

Tx   is T wave position

$$Tx = Sx + 1 \tag{10}$$

$$T \text{ wave} = (Tx, Ta) \tag{11}$$

where $Ta < |Ra|$

3.5   ECG Signal Classifications

An analysis of the ECG signal to classify Myocardial Infarction patients by detecting inverted T-wave will be conducted. If there is regularity in the T-wave that follows the QRS complex, the sample is healthy. But if there was an inverted T-wave following the QRS complex, the sample is Myocardial Infarction patient. The Myocardial Infarction classification algorithm is as follows.

```
If (Ta > 0) then Healthy Control
If (Ta < 0) then Myocardial Infarction end
```

where Ta is T wave amplitude.

4   Results and Discussion

The ECG signal data from the PTB Diagnostic ECG database is used to analyze ECG signal waveform. The samples are 10 healthy controls and 10 Myocardial Infarction patients, using 10 s of ECG signal in Lead I. The classification result of a Myocardial Infarction patient is presented in Table 1. This algorithm classified 7 correctly out of 10 healthy controls or 70% accuracy. The results showed that 30% of samples is false positive, this is acceptable because it does not harm the patient as physicians need to examine physically and verify laboratory results before final diagnose of a Myocardial Infarction patient.

Table 1. ECG analysis results

	Positive	Negative
True	8	7
False	3	2

For Myocardial Infarction patients, this algorithm classified 8 correctly out of 10 cases or 80%. The results show that 20% of samples is false negative, this is harmful as the patients are Myocardial Infarction, but the algorithm analyzed incorrectly. This would cause patients not to receive treatment in time. The overall result provides 75% accuracy of Myocardial Infarction patient classification, 80% sensitivity and 77.78% specificity.

5   Conclusions

Electrocardiogram (ECG) is a graph showing the heart electrical activity. The various characteristic features of ECG are used to classify the cardiac abnormalities and support decisions of the physicians. This paper purposes an ECG analysis algorithm using Wavelet transform to classify Myocardial Infarction patients. The steps in ECG signal analysis are noise elimination of ECG signal, R peak Detection, QRS Complex Detection and Myocardial Infarction Classification. In preprocessing stage, the noise eliminate, multilevel one-dimensional wavelet analysis uses a Daubechies family. Then, the wavelet decomposition of ECG signal is selected at level 4, using Daubechies4 since this waveform resembles the original ECG signal. In the feature extraction stage, a threshold base is applied in the algorithm for ECG signal classification. The analysis of ECG signals to classify Myocardial Infarction patient using Wavelet transform, the results show that an accuracy of Myocardial Infarction patient classification is 75%, with sensitivity at 80% and a specificity of 77.78%.

In future work, an enhanced technique will be applied for noise elimination to enhance the ECG signal quality. Moreover, variable values in the feature extraction algorithm will be adjusted to further improve the classification performance.

Acknowledgements. This research was funded by King Mongkut's University of Technology North Bangkok. Contract no. KMUTNB-GOV-59-50.

References

1. World Health Organization. Available from: http://www.who.int/mediacentre/factsheets/fs317/en/. cited 13 Oct 2014
2. Thygesen, K., et al.: Third Universal Definition of Myocardial Infarction. J. Am. Coll. Cardiol. **60**(16), 1581–1598 (2012)
3. Napodano, M., Paganelli, C.: ECG in Acute Myocardial Infarction in the Reperfusion Era. INTECH Open Access Publisher, (2012)
4. Martis, R.J., et al.: Cardiac decision making using higher order spectra. Biomed. Signal Process. Control **8**(2), 193–203 (2013)
5. Burns, N.: Cardiovascular physiology. Retrieved from School of Medicine, Trinity College, Dublin (2013). http://www.medicine.tcd.ie/physiology/assets/docs12_13/lecturenotes/NBurns/Trinity%20CVS%20lecture
6. Klabunde, R.: Cardiovascular Physiology Concepts. Lippincott Williams & Wilkins (2011)
7. Manikandan, M.S., Dandapat, S.: Wavelet-based electrocardiogram signal compression methods and their performances: A prospective review. Biomed. Signal Process. Control **14**, 73–107 (2014)
8. Acharya, R., et al.: Advances in Cardiac Signal Processing. Springer, New York (2007)
9. Saritha, C., Sukanya, V., Murthy, Y.N.: ECG signal analysis using wavelet transforms. Bulg. J. Phys **35**(1), 68–77 (2008)
10. Mateo, J., Sánchez-Morla, E.M., Santos, J.L.: A new method for removal of powerline interference in ECG and EEG recordings. Comput. Electr. Eng. (2015)
11. Pan, J., Tompkins, W.J.: A real-time QRS detection algorithm. IEEE Trans. Biomed. Eng. **3**, 230–236 (1985)
12. Martis, R.J., et al.: Application of principal component analysis to ECG signals for automated diagnosis of cardiac health. Expert Syst. Appl. **39**(14), 11792–11800 (2012)
13. Thomas, M., Das, M.K., Ari, S.: Automatic ECG arrhythmia classification using dual tree complex wavelet based features. AEU-Int. J. Electron. Commun. (2015)
14. Chang, P.-C., et al.: Myocardial infarction classification with multi-lead ECG using hidden Markov models and Gaussian mixture models. Appl. Soft Comput. **12**(10), 3165–3175 (2012)
15. Banerjee, S., Gupta, R., Mitra, M.: Delineation of ECG characteristic features using multiresolution wavelet analysis method. Measur. **45**(3), 474–487 (2012)

16. Sun, L., et al.: ECG analysis using multiple instance learning for myocardial infarction detection. IEEE Trans. Biomed. Eng. **59**(12), 3348–3356 (2012)
17. Banerjee, S., Mitra, M.: Application Of Cross Wavelet Transform For Ecg Pattern Analysis And Classification. (2014)
18. Goldberger, A.L., et al.: Physiobank, physiotoolkit, and physionet components of a new research resource for complex physiologic signals. Circ. **101**(23), e215–e220 (2000)
19. Zadeh, A.E., Khazaee, A., Ranaee, V.: Classification of the electrocardiogram signals using supervised classifiers and efficient features. Comput. Methods Programs Biomed. **99**(2), 179–194 (2010)
20. Manikandan, M.S., Soman, K.P.: A novel method for detecting R-peaks in electrocardiogram (ECG) signal. Biomed. Signal Process. Control **7**(2), 118–128 (2012)
21. Rai, H.M., Trivedi, A., Shukla, S.: ECG signal processing for abnormalities detection using multi-resolution wavelet transform and Artificial Neural Network classifier. Meas. **46**(9), 3238–3246 (2013)

The Algorithm of Static Hand Gesture Recognition using Rule-Based Classification

Thittaporn Ganokratanaa and Suree Pumrin[✉]

Department of Electrical Engineering, Chulalongkorn University, Bangkok,
Thailand
charisma_sbunny@hotmail.com, suree.p@chula.ac.th

Abstract. Technology becomes a part of human lives for decades, especially in human—computer interaction (HCI) that considered as the important research area involving with an assistive technology and a medical system. Hand gesture is classified as an intuitive method for human to interact with the computer. It is useful for elderly people who cannot express their feelings by words. This paper proposed the hand gesture recognition technique for the elderly by using contour detection with convex hull feature extraction and rule-based classification. Vision-based hand gesture recognition is considered in classifying six hand gestures as lingual description. From the experimental results, the hand gesture system provides good detection and classification results for all the six static hand gestures.

Keywords: Hand gesture recognition · Elderly · Contour detection · Convex hull · Rule-based · Vision-based

1 Introduction

Nowadays, technology is continuously and increasingly used in all the world. The computer technology has high effect and closely relationship to human. Many new devices are developed in the area of computer technology. One of multidisciplinary research areas is human-computer interaction (HCI) which is a critically important approach for the interaction between human and computer. It also creates other innovative technologies in various ways such as games, disaster and crisis management, human-robot interaction, assistive technology, and medical system.

In medical technology such as health care, HCI has an important role to help reduces patient's accidents by developing a monitoring system to track the physical activities and emotions [1]. This technology is useful for elderly or people with physical handicaps. From the older population in the United States [2], it has shown that the elderly will become the majority population in 2043. Therefore, the elderly should have been noticed more, especially when they are alone. For example, in the hospital where the caregivers have no time to take care one patient for a long time. The elders cannot express their requirements and situations such as toilet, water, and any accidents [3]. Thus, the use of monitoring system is helpful for both of the caregivers and elderly.

© Springer International Publishing AG 2018
T. Theeramunkong et al. (eds.), *Advances in Natural Language Processing,
Intelligent Informatics and Smart Technology*, Advances in Intelligent Systems

The monitoring system is the interaction language between human and computer. Hand gesture is an approach to the human to human and human to computer communications because it is based on the institution and considered as the natural gesture. Hand movements can convey information and notify the elderly thoughts [4]. Some applications applied in contact-based devices with markers and gloves required the elders to carry wires and connect to a device. From this solution, it causes uncomfortable feelings to the users. To solve this problem, the vision-base gesture algorithm is developed without the requirements of the devices attached to the elders. Thus, this paper focuses on the vision based technique by providing the hand gesture recognition techniques for the elderly.

The concept of this paper is to propose hand gesture recognition techniques for the elderly by using contour detection and convex hull. This paper is provided into five sections including, introduction in section one, preliminaries in section two, proposed method in section three, experimental results in section four, and conclusion in section five. In section two, researchers provide a survey of the related works of hand gesture recognition technologies. Section three shows the proposed method used for feature detecting, extracting, and classifying the hand gesture. Section four provides the experiment and experimental results. Finally, section five is to conclude this work.

2 Preliminaries

There are various hand gestures in static scene used to implement and recognition. The gestures should be simple, easy to remember and easy to communicate for both of the elderly and caregivers. The important fact is the gestures have to correspond with the lingual meaning. In this paper, six hand gestures which are easy to use in daily life are proposed and defined their meanings. The hand gestures are shown in Fig. 1. A static hand gesture is a posture of the hand without any movement. Figure 2a shows the Chinese sign language [6] which provides many postures to represent the numbers and alphabets and Fig. 2b shows example of gestures. Most of the static hand gesture researches request a fixed number for processing in various techniques. In [7], six gestures are presented with the descriptions.

Fig. 1. The proposed six hand gestures [5].

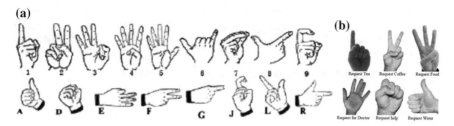

Fig. 2. Hand gestures; **a** The Chinese sign language [6] and **b** Example of hand gesture with lingual descriptions [7]

The key in vision based device is the development of a specific algorithm and the application of machine learning. There are various methods adapted to the hand gesture recognition such as Haar-like features and adaboost learning algorithm that are utilized for recognizing hand gesture [8]. The popular technique is partial contour matching proposed with skin color segmentation for extracting the silhouette of the hand and classifying the hand gesture in different posters [9]. Rule-based is also provided good results due to its simplicity and suitability for a small amount of data. According to related researches, the implementation of system can be concluded in these procedures. First, the image or video file is imported into the system in order to process the static or dynamic hand gesture as a real-time system. After that, splitting the video file to image frames and extracting features of hand to detect it as foreground. Feature extraction considers hand shape, size, and skin-color. The detection considers motion, color, shape, and depth of objects. The tracking processed by a template based correlation and contour based intensity. Finally, the recognition process includes a mean shift clustering applied with the particle filter, a hidden Markov model including a Bayesian estimator with a hidden non-Markov, a dynamic time warping, a support vector machine or multiclass SVM. However, these techniques require a large data and sometimes lead to high runtime complexity. For this paper, six hand gestures are classified as the common command for practical use. To find the line around the hand and point of convexity defect to locate the fingers, this paper focuses on the hand shape detection by using contour and convex hull and classifies the hand gesture with rule-based algorithm.

3 Proposed Method

In the process of the hand gesture recognition, the suitable techniques are employed to implement and classify the hand gesture in the system. The process is divided into seven procedures, including video receiving, image capturing, background subtraction, image moments, contour detection, convex extraction, and rule-based classification as shown in Fig. 3

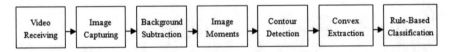

Fig. 3. Procedures of the hand gesture recognition system

3.1   Video Receiving

This is the first step in the hand gesture recognition system requiring the image frames from a video stream file saved in a form of .avi. It works in real time by importing the video file and then splitting into image frames as explained in the image capturing process. The image frames will be the same size as the resolution of the monitoring camera without any additional configuration.

3.2   Image Capturing

This process analyzes the video file to apply in splitting image frames. The system implements the data in the video file stored in the stacked array. Then it specifies the variable in order to divide sub-frames of the variable in the image array to the image frames. After dividing all the frames, the system collects all the frames in the frame buffer that displayed in the storage module for analysis in the next process.

3.3   Background Subtraction

This is a useful technique applied with the static camera before implementing the hand gesture in the feature extraction and classification processes. It can extract the human hand considered as the foreground object by indicating the difference between the current frame and reference frame. It removes the background image to implement with the pixel of the hand. The background subtraction begins with the conversion of RGB color image to gray scale image. Then, it applies Gaussian to filter noise signal for improving image quality and finally determines Threshold to convert gray scale to binary image, in which white pixels represent hand region and black pixels represent other regions. The equation of the background subtraction is

$$o_{ij} = c_{ij} - r_{ij}, \tag{1}$$

where o_{ij} is the output result between the current frame and reference frame, c_{ij} is the current frame, and r_{ij} is the reference frame.

3.4   Image Moments

Image moments show the average of the intensity of pixels according to x axis and y axis. A set of clustering values is characterized by a few numbers related to its moments and the sums of integer powers [10]. Moment features implemented without considering the image location and size of the hand. The function $M(i,j)$ defines the hand in an image and generates the moments M_{ij} to provide features of the hand as shown in Eqs. (2) and (3). Image moments consist of centre of mass, variance and orientation as described as follows

$$M_i = \sum_{ij} x_i M(i,j), \tag{2}$$

$$M_j = \sum_{ij} y_j M(i,j). \tag{3}$$

Centre of mass. It is a center of area called centroid defined by the intensity of $M(i, j)$ at each point (i, j) of the given image (M) that is represented as the mass of (i, j). The object mass can be concentrated regardless of changing the first moment of the hand at any axis. The centre of mass can be calculated by the Eqs. (4) and (5) as follows

$$M_{10} = \sum \sum x_i M(i,j)/M_{00}, \tag{4}$$

$$M_{01} = \sum_{ij} \sum y_j M(i,j)/M_{00}. \tag{5}$$

Variance. (σ^2) It is obtained from the second moment of the centroid (M_{ij}^a). Variance is calculated as

$$\sigma_x^2 = M_{20}^a = M_{20} - M_{10}^2, \tag{6}$$

$$\sigma_y^2 = M_{02}^a = M_{02} - M_{01}^2. \tag{7}$$

Orientation. It represents the angle of axis of the least moment of the inertia. The calculation is shown in Eqs. (8), (9), and (10)

$$\tan(2\theta) = \frac{2M_{11}}{M_{20} - M_{02}}, \tag{8}$$

unless $M_{11} = 0$ and $M_{20} = M_{02}$ consequently,

$$\sin(2\theta) = \frac{\pm 2M_{11}}{\sqrt{(2M_{11})^2 + (M_{20} - M_{02})^2}}, \tag{9}$$

$$\cos(2\theta) = \frac{\pm(M_{20} - M_{02})}{\sqrt{(2M_{11})^2 + (M_{20} - M_{02})^2}}. \tag{10}$$

From these equations, it was found that only the first and second moments are required to calculate the centre of mass, variance and orientation of the hand. In linear transformation, the moments are invariant meaning that the original image is not necessary to obtain the first and second moments. After processing these steps, the centre of mass of the hand is found and used to apply in the next step.

3.5   Contour Detection

Contour means a curve used to connect the pixels with the same intensity. It illustrated by contour lines which are possible to have more than one line in one image. The contour presents the steepness of slopes and draws around the white blob of the hand which is determined by threshold value [11]. The blob will be performed more than one blob in the image depends on the noise in the image. If the noises are small, the blobs will present the large contour of the hand. The green outline represents the largest contour of the hand which is shown in Fig. 4. In modeling, the contour is defined in the (x, y) plane of an image as a parametric curve shown as a follow equation:

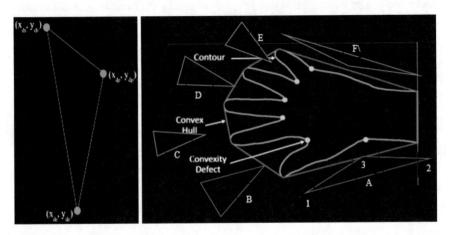

Fig. 4. The illustration of contour, convex hull, and convexity defect with the six defect triangles; A, B, C, D, E, F, and the three coordinates; the defect start (x_{ds}, y_{ds}), defect end (x_{de}, y_{de}), and defect position points (x_{dp}, y_{dp}), labeled as 1, 2, and 3, respectively

$$v(s) = (x(s), y(s)). \tag{11}$$

Contour provides energy (e_{snake}) defined as the sum of the three energy terms including, e_{int}, e_{ext}, and e_{cons}, which represent internal, external and constraint, respectively. The energy is calculated as

$$e_{snake} = e_{int} + e_{ext} + e_{cons}. \tag{12}$$

From the above energy terms, it was found that the final position of the contour has a minimum energy (e_{min}). In an internal energy (e_{int}), it is derived from the sum of the elastic energy (e_{elas}) and bending energy (e_{bend}) which are described below:

Elastic energy represents an elastic potential energy that decreases stretching and suits for shrinking the contour. The elastic energy is provided in the Eq. (13)

$$e_{elas} = \frac{1}{2}\int_S \alpha(s)|v_s|^2\, ds, \tag{13}$$

where weight is presented as $\alpha(s)$ for controlling the elastic energy with different parts of contour. Bending Energy (e_{bend}) applied for a thin metal strip. It is calculated by the sum of squared curve of the contour. The equation of the bending energy is shown as

$$e_{bend} = \frac{1}{2}\int_S \beta(s)|v_{ss}|^2\, ds, \tag{14}$$

where $\beta(s)$ is served as similar role to $\alpha(s)$. Thus, the internal energy can be calculated as

$$e_{int} = e_{elas} + e_{bend} = \int_S \frac{1}{2}(\alpha|v_s|^2 + \beta|v_{ss}|^2)ds. \tag{15}$$

For finding an external energy, the external energy implements on small values, including boundaries. It can be derived from the image defined by a function $e_{img}\,(x, y)$,

$$e_{ext} = \int_s e_{img}(v(s))ds. \tag{16}$$

Thus, the energy (e_{snake}) equation which can find a contour $v(s)$ can be calculated to minimize the energy function of the hand by Lagrange equation as shown in following equation:

$$e_{snake} = \int_S \frac{1}{2}(\alpha(s)|v_s|^2 + \beta(s)|v_{ss}|^2) + e_{img}(v(s))ds, \tag{17}$$

$$\alpha v_{ss} - \beta v_{ssss} - \nabla e_{img} = 0. \tag{18}$$

3.6 Convex Extraction

Convex extraction is utilized as the feature extraction which includes convex hull and convexity defects described as follow:

Convex hull. This is the largest contour provided without the curve of the hand. It represents convex shape around the hand contour [12]. The red line in Fig. 4 shows the convex hull of the detected hand. The convex hull can be divided into three parts, including lines, segments, and polygons as explained below:

Lines can be represented as l performs a triple (a, b, c), in which a, b, and c are the coefficients of the linear equation

$$ax + by + c = 0. \tag{19}$$

Moreover, two different points such as q_1 and q_2, can defined and connected with the line l by providing the Cartesian coordinates to q_1 and q_2, including (x_1, y_1) and (x_2, y_2) of q_1 and q_2, respectively. The line l through q_1 and q_2 is given by the following equation

$$\frac{x - x_1}{x_2 - x_1} = \frac{y - y_1}{y_2 - y_1}, \tag{20}$$

then we derive

$$a = (y_2 - y_1), \tag{21}$$

$$b = -(x_2 - x_1), \tag{22}$$

$$c = y_1(x_2 - x_1) - x_1(y_2 - y_1), \tag{23}$$

where segment s is represented by giving the pair (p, q) of points in the (x, y) plane that form the endpoints of s and giving the line through the points with a range of xy coordinates that restricted to s and polygon P is represented by performing a circular sequence of points called the vertices of P. The edges of P are the segments between P consecutive vertices. Polygon is considered as convex when it is simple shape, nonintersecting line, and all its angles are less than π.

Convexity defect. After drawing the convex hull around the contour line of the hand, the contour points are defined within the hull by using minimum points. The formation of defects is provided in the convex hull due to the contour of the hand. A defect is at the hand contour away from the convex hull. The set of values for every defect in the convexity is formed as vector. This vector contains the start, end, and defect points of the line in the convex hull. These points indicate the coordinate points of the contour [13]. Figure 4 shows the convexity defects represented in yellow point.

In classification, the feature vectors are derived from the hull defects. The convexity defect in the hull is the distance between the contour line and actual hand. When the camera monitors the hand, the three coordinate points of defect, including the start (x_{ds}, y_{ds}), end (x_{de}, y_{de}), and position points (x_{dp}, y_{dp}), labeled as 1, 2, and 3, respectively, can be described by six defect triangles (A, B, C, D, E, F) in each frame as shown in Fig. 4. Thus, defect triangle A can be described as the following vector, called V_{td}^A, including [A (x_{ds}) A (y_{ds}) A (x_{dp}) A (y_{dp}) A (x_{de}) A (y_{de})].

From the convex defects, five largest defect areas of the hand is defined based on the defect position points (x_{dp}, y_{dp}) from the top six triangles: A, B, C, D, E, F. Accurate registration is important between the frames for further analysis.

The distance between two pixels. In x and y coordinates that restricted to s, we assume that the distance d can be calculated by defining the value of pixel one at the (i_1, j_1) and pixel two at the (i_2, j_2) as shown in Eq. (24)

$$d = \sqrt{(i_2 - i_1)^2 + (j_2 - j_1)^2}. \tag{24}$$

3.7 Rule-Based Classification

Rule-based classification provides a set of encoded rules extracted from input gestures and compared with feature inputs. The output gesture is shown by matching the input gestures and the rules [14]. The rules consider the distance between the centroid of the hand and any point of fingertips. The threshold is set at 200 pixels. If the distance is less than 200 pixels, it will classify the gesture as stop. Conversely, if the distance is more than 200 pixels, it will classify the gesture according to the number of lines that occurred. The rules are divided into six procedures as shown in Fig. 5.

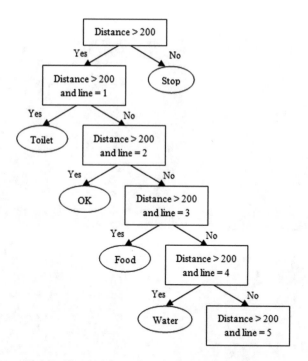

Fig. 5. The decision rules of rule-based classification

From Fig. 5, it shows the decision rules that can be described below:

Rule 1: The system interprets the lingual description of the close fist image that has no straight line longer than 200 pixels from the centroid of the hand drawn to any point of fingertips as "stop".

Rule 2: The system interprets the lingual description of the pointing finger image that has a straight line longer than 200 pixels drawn from the centroid of the hand towards the pointing fingertip as the requirement of "toilet".

Rule 3: The system interprets the lingual description of the image of pointing and middle fingers that has two straight lines longer than 200 pixels drawn from the centroid of the hand towards the pointing and middle fingertips as "ok".

Rule 4: The system interprets the lingual description of the image of pointing, middle, and forth fingers that has three straight lines longer than 200 pixels drawn from the centroid of the hand towards the pointing, middle, and fourth fingertips as the requirement of "food".

Rule 5: The system interprets the lingual description of the image of pointing, middle, forth, and little fingers that has four straight lines longer than 200 pixels drawn from the centroid of the hand towards the pointing, middle, fourth, and little fingertips as the requirement of "water".

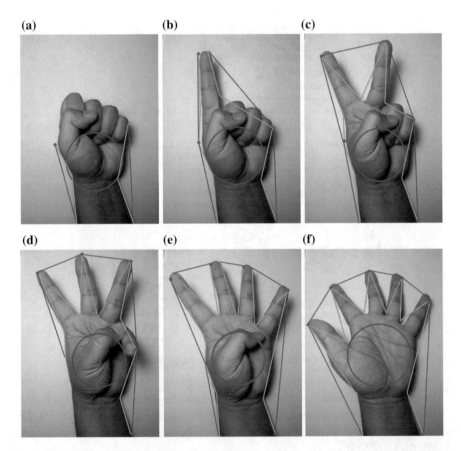

Fig. 6. The experimental results of the recognition of the six hand gestures, including **a** the close fist represents "stop", **b** the pointing finger represents "toilet", **c** the pointing and middle fingers represent "ok", **d** the pointing, middle, and forth fingers represent "food", **e** the pointing, middle, forth, and little fingers represent "water", and **f** the open palm represents "help"

Rule 6: The system interprets the lingual description of the open palm image that has five straight lines longer than 200 pixels drawn from the centroid of the hand towards the five fingertips as the requirement of "help".

4   Experimental Results

The hand gesture recognition system is implemented by analyzing the images of the six different static hand gestures under the closed environment. The camera is set from the top level above the hand about 30 cm. A static white paper placed at the reference level to be used as a scene in the background. The light source comes from fluorescent bulb that was about two meters above the reference level. The results show the good performance of classification as follows:

From the experimental results shown in Fig. 6, the hand gesture recognition system provides good results in hand detection and classification for all the six hand gestures. The centroid and contour line of the hand and the points of the fingertips are detected precisely. The hand gestures indicate their meaning well, including the close fist represents "stop", the pointing finger represents "toilet", the pointing and middle fingers represent "ok", the pointing, middle, and forth fingers represent "food", the pointing, middle, forth, and little fingers represent "water", and the open palm represents "help".

5   Conclusions

This paper proposed the hand gesture recognition techniques applied for the elderly. The process consists of three main parts, including detection, feature extraction, and classification using contour, convex hull, and rule-based algorithms, respectively. The system focuses on vision-based hand gesture recognition of the six static hand gestures. From the experimental results, the hand gesture recognition system provides good results in detecting and classifying all the six hand gestures as the lingual descriptor. The results depend on some conditions such as the static background, the distance between the camera and hand, and the light conditions.

Acknowledgement. This research has been supported by Embedded System and IC Design research laboratory, Department of Electrical Engineering and Chulalongkorn Academic Advancement into Its 2nd Century Project. The Scholarship from the Graduate School, Chulalongkorn University to commemorate the 72nd anniversary of his Majesty King Bhumibala Aduladeja is also gratefully acknowledge.

References

1. Acharya, C., Thimbleby, H., Oladimeji, P.: Human computer interaction and medical devices. In: the 24th BCS Interaction Specialist Group Conference, pp. 168–176. British Computer Society (2010)
2. Ortman, J.M., Velkoff, V.A., Hogan, H.: An aging nation: the older population in the United States, pp. 25–1140. US Census Bureau, Washington, DC (2014)

3. Chaudhary, A., Raheja, J.L.: A health monitoring system for elder and sick persons. *arXiv preprint* arXiv:1304.4652 (2013)
4. Rautaray, S.S., Agrawal, A.: Vision based hand gesture recognition for human computer interaction: a survey. Artif. Intell. Rev. **43**(1), 1–54 (2015)
5. Jalab, H.A.: Static hand Gesture recognition for human computer interaction. Inf. Technol. J. **11**(9), 1265–1271 (2012)
6. Chaudhary, A., Raheja, J.L., Das, K., Raheja, S.: Intelligent approaches to interact with machines using hand gesture recognition in natural way: a survey. *arXiv preprint* arXiv: 1303.2292 (2013)
7. Chaudhary, A., Raheja, J.L.: A health monitoring system for elder and sick persons. *arXiv preprint* arXiv:1304.4652 (2013)
8. Chen, Y.T., Tseng, K.T.: Developing a multiple-angle hand gesture recognition system for human machine interactions. In: 33rd annual conference of the IEEE industrial electronics society (IECON), pp 489–492 (2007)
9. Ohn-Bar, E., Trivedi, M.M.: Hand gesture recognition in real time for automotive interfaces: A multimodal vision-based approach and evaluations. Intell. Transp. Syst. IEEE Trans. **15**(6), 2368–2377 (2014)
10. Chonbodeechalermroong, A., Chalidabhongse, T.H.: Dynamic contour matching for hand gesture recognition from monocular image. In: 2015 12th International Joint Conference on Computer Science and Software Engineering (JCSSE), pp. 47–51. IEEE (2015)
11. Peterfreund, N.: Robust tracking of position and velocity with Kalman snakes. Pattern Anal. Mach. Intell. IEEE Trans. **21**(6), 564–569 (1999)
12. Goodrich, M.T., Tamassia, R.: Algorithm design: foundation, analysis and internet examples. Wiley (2006)
13. Youssef, M.M.: Hull convexity defect features for human action recognition. University of Dayton, Diss (2011)
14. Ren, Z., Yuan, J., Meng, J., Zhang, Z.: Robust part-based hand gesture recognition using kinect sensor. In: IEEE Transactions on Multimedia, vol. 15, no.5 (2013)

Robust Watermarking for Medical Image Authentication Based on DWT with QR Codes in Telemedicine

Narit Hnoohom[1], Chanathip Sriyapai[2], and Mahasak Ketcham[2(✉)]

[1] Image, Information and Intelligence Laboratory, Department of Computer Engineering, Faculty of Engineering, Mahidol University, 25/25, Phuttamonthon 4 Road, Salaya 73170, Nakhon Pathom, Thailand
narit.hno@mahidol.ac.th
[2] Department of Information Technology Management, Faculty of Information Technology, King Mongkut's University of Technology North Bangkok, 1518, Pracharat 1 Road, Wongsawang, Bangsue, 10800 Bangkok, Thailand
s5707021857106@email.kmutnb.ac.th, mahasak.k@it.kmutnb.ac.th

Abstract. Recently, the exchange of medical images over the internet in order for diagnosis, treatment and telemedicine consulting in remote areas has become more important. As a result of information exchanging, and as the data in question is sensitive, watermark embedding is then used in the medical image processing in order to prevent the digital content from being modified without permission as well as to verify the authenticity of medical images. In this paper, Discrete Wavelet Transform (DWT) based image watermarking method are used to hide patient's data into the Region of Interest (ROI) to show the identity and to verify the data authenticity. The contents of ROI is important for diagnostic decision making. Thus, in this paper, the researchers give priority to the imperceptibility performance. Patient's data will be taken to create a Quick Response (QR) code before embedding watermarks in order to increase the security level of data and the data capacity. The experiment's results reveal that imperceptibility performance can be achieved. This watermark embedding approach is high in the Peak Signal-to-Noise Ratio (PSNR) values and robust from various attacks. The extracted watermark can be decoded back into text information although it was attacked.

Keywords: Digital watermarking · Medical image · Telemedicine · Discrete wavelet transform · Quick response code · Region of interest

1 Introduction

Telemedicine is the use of telecommunication and information technologies in order to provide clinical healthcare at a distance by medical professionals and it is useful in helping to diagnose the treatment and prevention of diseases. Beneficially, it facilitates medical education to medical personnel and especially to the store-and-forward tele-medicine (Asynchronous) which receives and transmits medical information. In general, physicians will not be able to get patient's information or perform a physical

© Springer International Publishing AG 2018
T. Theeramunkong et al. (eds.), *Advances in Natural Language Processing, Intelligent Informatics and Smart Technology*, Advances in Intelligent Systems

examination from the source directly, therefore, they must rely on medical history and data images or videos that have been passed down [1]. When the medical images and patient's data are exchanged together an advanced security must be in operation in order to address the medical data issues which have been trans-mitted through the communicating network. This may affect the image and data including the confidentiality of patient's information or data may be damaged during transit and errors may appear on the image as a result. In addition, if there is any mistake because of the unexpected patient's information inter-change, it is possible to lead to incorrect treatment. From the study, [2] it states that application the digital watermark embedding for the telemedicine on medical images can meet two main issues; safety, which is used to control and verify image authenticity, and the saving of bandwidth from sending data and embedding useful descriptions such as patient's electronic records.

From the study of research and the experimental development for embedding watermarks on medical images, two dived groups of Spatial Domain and Transform Domain techniques are provided;

The first group, [3, 4] before embedding fragile watermarks into medical images, must find the ROI and the Region of Non Interest (RONI) in order to keep the value of data recovery if the image is unexpectedly attacked or damaged. It is mainly noted to avoid embedding watermarks in ROI to preserve the image quality and embed watermarks in RONI instead. However, the disadvantage of this technique is that if the RONI has been damaged, it is unable to be retrieved even through [5] an embedded robust watermark.

The second group, [6–15] embedding permanent/irreversible watermarks into medical images is generally embedded into the whole image by using a Transform Domain technique which is more effective than the Spatial Domain and does not affect the image quality as much.

In addition, the research [16, 17] analyses hiding the QR codes with various techniques. It was found that embedding watermarks using DWT achieves high imperceptibility and the use of QR codes can increase the capacity of hiding information as well.

In this paper, the author proposes to embed watermarks into ROI for medical images, where watermarking is specifically related to confidentiality, high resistance to attack, and causes the lowest negative result on image quality. In this paper, we give priority to imperceptibility performance because the contents of ROI are important for diagnostic decision making and authentication by using a specific technique for hiding patient's data and a QR code to increase the data capacity as well as to protect patient's data confidentiality.

2   Proposed Method

The objective of the proposed method is to hide patient's data into the ROI to show the identity and to verify the data authenticity. A block diagram of the proposed method is shown in Fig. 1.

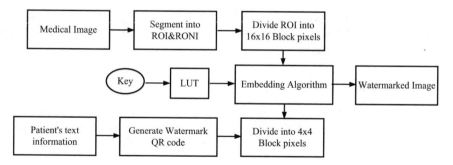

Fig. 1. Block diagram of proposed method

2.1 Segmentation of Medical Images

In the proposed method, medical images are a different size to grayscale images. The medical image is segmented into two regions: ROI and RONI pixels. In a medical image, the ROI area is generally marked by a physician interactively and is in any irregular shape as shown in Fig. 2.

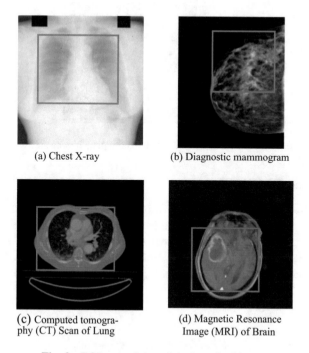

Fig. 2. ROI area of the original medical image

A medical image may contain a number of ROI areas [5]. In this paper, medical images with one ROI area are considered. The size of the ROI in this experiment are defined as 512×512 pixels.

2.2   Divide ROI into Block Pixels

From the medical image, divide ROI area into non-overlapping blocks size 8×8 pixels. The researchers define X is the ROI area of the original medical image which is a greyscale image size $N_1 \times N_2$ pixels. It is represented by Eq. (1).

$$X = \{ X(i,j);\ 0 \leq i < N_1,\ 0 \leq j < N_2 \} \tag{1}$$

where $X(i,j) \in \{0, \ldots, 2L - 1\}$ is the intensity of pixel at (i, j) and L is the number of bits which is in bit in each pixel.

The amount of all block pixels are associated with the size of the original image ROI as $\frac{N_1}{8} \times \frac{N_2}{8}$ block pixels or 64×64 blocks when $N_1 = N_2 = 512$ as shown in Fig. 3.

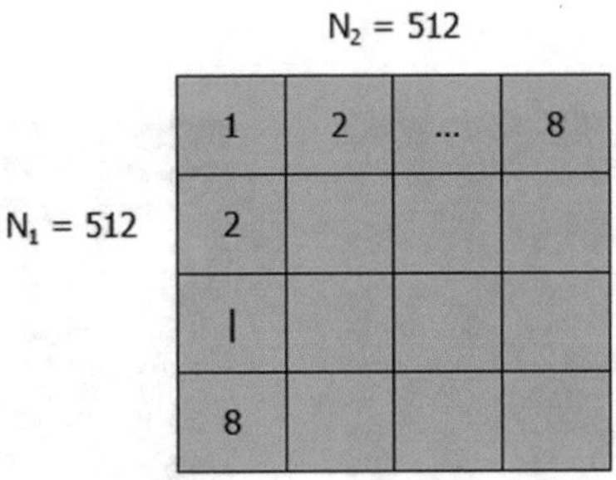

Fig. 3. The ROI area is divided into non-overlapping blocks size 8×8 pixels

2.3   Generating Watermark Image

In the proposed method, the watermark in this experiment is QR code image. It was generated from the text watermark which is simulated clinical data.

Figure 4 shows the contents of the text watermark. Clinical data and information are arranged in the order of ID number, first name, last name, age, sex, the date the picture was created and notes for remarks. Figure 5 shows the QR code Image which is generated from the text watermark.

HPT00123 chana sriyapai
26 m 16012015 MRI

Fig. 4. Text watermark

Fig. 5. Watermark image

The QR code image is a binary image size 64 × 64 pixels. The binary image comes with two different levels. Black and white with the values 0 and 1 respectively. The QR code version that the researchers used in this experiment is version 4 (Modules 33 × 33). The error correcting function of the QR code for misreading is defined as level Q (25%).

2.4   Divide Watermark into Block Pixels

The watermark image, which is a binary QR code image divided into non-overlapping blocks size 1 × 1 pixels. The researchers define W as the Watermark image which is binary image size $M_1 \times M_2$ pixels. There are only two levels of intensity in the binary image. Black and white with the values 0 and 1 respectively. It is represented by Eq. (2)

$$W = \{W(i, j); 0 \leq i < M_1, 0 \leq j < M_2 \tag{2}$$

where $W(i, j) \in \{0, 1\}$ for I, j where $0 \leq i < M_1, 0 \leq j < M_2$.

Size of each block of watermark is associated with the original medical image and watermark image as $\left[M_1 \times \frac{8}{N_1} \right] \times \left[M_2 \times \frac{8}{N_2} \right]$ block pixels.

In this paper, the ROI area size 512 × 512 pixels and watermark image size 64 × 64 pixels then, size of each block of watermark is 1 × 1 pixels as shown in Fig. 6.

M₂ = 64 → $M_2 = 64$

	1	2	...	64
2				
I				
64				

$M_1 = 64$

Fig. 6. The watermark image is divided into non-overlapping blocks size 1 × 1 pixels

2.5   Embedding Algorithm

Step 1 For each block of ROI area (X) transformed into frequency domain performed by Haar Discrete Wavelet Transform with 3-level decomposition for embedding the watermark image in (LH^3) and (HL^3) as shown in Fig. 7.

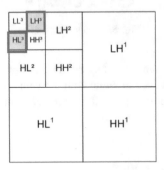

Fig. 7. The DWT with 3-level decomposition

After performing the wavelet transformation on each block of the ROI area size 8×8 pixels, size of block pixels will be reduced to the size of a digital watermark 1×1 pixels.

Step 2 Embedding the watermark image (W) into the ROI area of the medical image (X) by changing the value of coefficient in X using Look-up Table (LUT) [18]. LUT was created by pseudorandom numbers using a secret key. The embedding algorithm can be defined as Eq. (3).

$$x_i' = \begin{cases} x_i & \text{if } LUT(x_i) = w_i \\ x_i + \delta & \text{if } LUT(x_i) \neq w_i \text{ and} \\ & \delta = \min_{|d|}\{d = x_t' - x_t : LUT(x_t') = w_t\} \end{cases} \quad (3)$$

Where

x_i Coefficient value of ROI area of the medical image (X)
x_i' Coefficient value of watermarked ROI area
w_i Coefficient value of watermark image (W)
$LUT(\cdot)$ Look-up Table.

Step 3 Transform the Domain which in frequency domain back into Time Domain by Inversing the Discrete Wavelet Transform for all block pixels.
Step 4 Merge all block pixels back to the ROI area.
Step 5 Merge the watermarked ROI area into the medical image. Finally, the researchers will get the watermarked medical image.

3   Experiment's Result

In the experiment, medical images were taken from the online database [19, 20] for embedded watermarks. The metric used in the experiment that we conducted was Peak Signal-to-Noise Ratio (PSNR) for measuring the quality of watermarked medical images that are defined as Eq. (4). The PSNR is the ratio between the reference signal and the distortion signal in an image. The higher the PSNR, the closer the distorted image is to the original. In general, a higher PSNR value should correlate to a higher quality image.

$$PSNR = 10 \log_{10}\left(\frac{255^2}{MSE}\right) dB \tag{4}$$

For MSE (Mean Squared Error) can define as Eq. (5).

$$MSE = \frac{\sum (f_w(x,y) - f(x,y))^2}{n} \tag{5}$$

Mean Squared Error is the average squared difference between a reference image and a distorted image. Where $f_w(x,y)$ denotes the watermarked medical image, $f(x,y)$ denotes the original medical image, and n denotes the number of pixels for the watermarked medical image and original medical image.

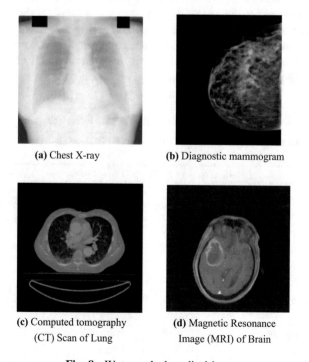

(a) Chest X-ray   (b) Diagnostic mammogram

(c) Computed tomography   (d) Magnetic Resonance
   (CT) Scan of Lung   Image (MRI) of Brain

Fig. 8. Watermarked medical image

The experiment was conducted on 10 medical images of X-ray modality, 10 medical images of Mammogram Modality, 10 medical images of CT scan modality and 10 medical images of magnetic resonance imaging (MRI) scan modality. Figure 8 shows a few medical images which were used for this experiment. All these images are 8 bits grayscale medical images (Fig. 9).

 HPT00123 chana sriyapai 26 m 16012015 Xray

(a) QR codefor X-ray

 HPT00123 chana sriyapai 26 m 16012015 Mamogram

(b) QR code for Mammogram

 HPT00123 chana sriyapai 26 m 16012015 CT

(c) QR code for CT scan

 HPT00123 chana sriyapai 26 m 16012015 MRI

(d) QR code for MRI scan

Fig. 9. Extracted watermark image

In order to evaluate the extracted watermark image, the researchers employ the normalized correlation (NC) [18] to measure the similarity between the original watermark and the extracted watermark. The NC value lies in the range [0, 1]. The larger NC value means the extracted watermark is completely restored. The contents of the ROI are important for diagnostic decision making. Therefore, the researchers give priority to imperceptibility performance. From Tables 1 and 2, it can be seen that this paper's watermarking approach has a high imperceptibility performance (high value of PSNR). The embedding of the QR code binary image into the ROI of Medical Images can contain more text information as well.

Table 1. Detail of medical images of diferent modalities and values of PSNR

Modality of medical image	Size of image	ROI pixels X	ROI pixels Y	MSE	PSNR	NC
X-ray	2048 × 2048	1060–1571	483–994	0.3724	52.4206	1
Mammogram	1422 × 1703	551–1062	734–1245	0.5176	50.9913	1
CT scan	1024 × 1024	274–785	208–719	0.3239	53.0272	1
MRI scan	1024 × 1024	241–752	365–876	0.4244	51.8536	1

Table 2. Average performance of the proposed method

Modality of medical image	Average MSE	Average PSNR	Average NC
X-ray	0.3391	52.8977	1
Mammogram	0.3716	52.2325	1
CT scan	0.4084	52.2401	1
MRI scan	0.3780	52.3035	1

Table 3. Result of applying various types of attacks on the watermarked medical images

Medical image	Types of attacks	NC	Extracted	Decoded text watermark
	Salt & pepper (density = 0.0005)	0.9841		HPT00123 chana sriyapai 26 m 16012015 Xray
	Salt & pepper (density = 0.001)	0.9688		HPT00123 chana sriyapai 26 m 16012015 Xray
	Rotation 90°	0.5796		HPT00123 chana sriyapai 26 m 16012015 Xray
	Rotation 180°	0.5625		HPT00123 chana sriyapai 26 m 16012015 Xray
	Image modification	0.9810		HPT00123 chana sriyapai 26 m 16012015 Xray
	Salt & pepper (density = 0.0005)	0.9832		HPT00123 chana sriyapai 26 m 16012015 Mamogram
	Salt & pepper (density = 0.001)	0.9734		HPT00123 chana sriyapai 26 m 16012015 Mamogram
	Rotation 90°	0.6025		HPT00123 chana sriyapai 26 m 16012015 Mamogram
	Rotation 180°	0.5381		HPT00123 chana sriyapai 26 m 16012015 Mamogram
	Image modification	0.9709		HPT00123 chana sriyapai 26 m 16012015 Mamogram
	Salt & pepper (density = 0.0005)	0.9875		HPT00123 chana sriyapai 26 m 16012015 CT
	Salt & pepper (density = 0.001)	0.9739		HPT00123 chana sriyapai 26 m 16012015 CT
	Rotation 90°	0.5669		HPT00123 chana sriyapai 26 m 16012015 CT
	Rotation 180°	0.6074		HPT00123 chana sriyapai 26 m 16012015 CT
	Image modification	0.9753		HPT00123 chana sriyapai 26 m 16012015 CT
	Salt & pepper (density = 0.0005)	0.9822		HPT00123 chana sriyapai 26 m 16012015 MRI
	Salt & pepper (density = 0.001)	0.9695		HPT00123 chana sriyapai 26 m 16012015 MRI
	Rotation 90°	0.5752		HPT00123 chana sriyapai 26 m 16012015 MRI
	Rotation 180°	0.5950		HPT00123 chana sriyapai 26 m 16012015 MRI
	Image modification	0.9778		HPT00123 chana sriyapai 26 m 16012015 MRI

From Table 3, it can be seen that this paper's watermarking approach has a high robustness for noise pollution and image rotation. The advantages of our proposed method is that patient's information can contain more detailed text information and the content of the extracted QR code image can be decoded normally.

4 Conclusions

In this paper the researchers proposed a method for clinical and patient's data hiding by applying the QR code technique. The embedding involved the DWT technique and Look-up Table to embed the QR codes image into ROI of Medical Images. The experiment's results reveal that imperceptibility performance can be achieved. Our watermark information can relatively contain longer text information. It is robust from various attacks and can be decoded from 2D barcode back into text. In any future work, the process of segmentation of medical images should be developed to detect the ROI area automatically and the robustness performance should be improved for other types of attacks as well.

References

1. Pharm, D., La-ongkaew, S.: [Online]. Telemedicine. Source: http://www.bangkokhealth.com/index.php/health/health-general/general-health/2146-telemedicine.html (16 March 2012)
2. Hussain, N., Boles, W., Boyd, C.: A review of medical image watermarking requirements for teleradiology. J. Digit. Imaging **26**(2), 326–343 (2013)
3. Eswaraiah, R., Edara, S.R.: ROI-based fragile medical image watermarking technique for tamper detection and recovery using variance. In: 2014 Seventh International Conference on Contemporary Computing (IC3), IEEE (2014)
4. Eswaraiah, R., Edara, S.R.: A fragile ROI-based medical image watermarking technique with tamper detection and recovery. In: Proceedings of the 2014 fourth international conference on communication systems and network technologies. IEEE Computer Society (2014)
5. Eswaraiah, R., Edara, S.R.: Robust medical image watermarking technique for accurate detection of tampers inside region of interest and recovering original region of interest. Image Process. IET **9**(8), 615–625 (2015)
6. Moniruzzaman, M., Hawlader, K.A.M., Hossain, F.M.: Wavelet based watermarking approach of hiding patient information in medical image for medical image authentication. In: 2014 17th International Conference on Computer and Information Technology (ICCIT), IEEE (2014)
7. Yatindra, P., Dehariya, S.: A more secure transmission of medical images by two label DWT and SVD based watermarking technique. In: 2014 International Conference on Advances in Engineering and Technology Research (ICAETR), IEEE (2014)
8. Nassiri, B., Latif, R., Toumanari, A., Maoulainine, F.M.R.: Secure transmission of medical images by watermarking technique. In: 2012 International Conference on Complex Systems (ICCS), IEEE (2012)

9. Dong, C., Zhang, H., Li, J., Chen, Y.-W.: Robust zero-watermarking for medical image based on DCT. In: 2011 6th International Conference on. IEEE Computer Sciences and Convergence Information Technology (ICCIT) (2011)

10. Ketcham, M., Ganokratanaa, T.: The evolutionary computation video watermarking using quick response code based on discrete multiwavelet transformation. In: Recent Advances in Information and Communication Technology. , pp. 113–123. Springer International Publishing (2014)

11. Ketcham, M.: The robust audio watermarking using singular value decomposition and adaptive tabu search. J. Convergence Inf. Technol. 9(3), 155 (2014)

12. Ketcham, M., Vongprahip, S.: An algorithm for intelligent audio watermarking using genetic algorithm. IEEE Congress on Evolutionary Computation, 2007, CEC 2007. IEEE (2007)

13. Vongpraphip, S., Ketcham, M.: An intelligence audio watermarking based on DWT-SVD using ATS. WRI Global Congress on Intelligent Systems, 2009, GCIS'09, Vol. 3. IEEE (2009)

14. Ketcham, M., Vongpradhip, S.: Intelligent audio watermarking using genetic algorithm in DWT domain. Int. J. Intell. Technol. 2(2), 135–140 (2007)

15. Ketcham, M., Vongpraphip S.: Genetic algorithm audio watermarking using multiple image-based watermarks. In: International Symposium on Communications and Information Technologies, 2007, ISCIT'07. IEEE (2007)

16. Rungraungsilp, S., Ketcham, M., Kosolvijak, V., Vongpradhip, S.: Data hiding method for QR code based on watermark by compare DCT with DFT domain. In: 3rd international conference on computer and communication technologies, India, 2012, Communication Technologies 2012, pp. 144–148. (2012)

17. Vongpradhip, S., Rungraungsilp, S.: QR code using invisible watermarking in frequency domain. In: 2011 9th International Conference on ICT and Knowledge Engineering (ICT & Knowledge Engineering), IEEE (2012)

18. Hnoohom, N., Vongpradhip, S.: Fragile watermarking based on look-up table. In: 1st Joint International Information Communication Technology, Lao PDR (2007)

19. Japanese Society of Radiological Technology (JSRT). [Online]. Digital image database. Source: http://www.jsrt.or.jp/jsrt-db/eng.php

20. The Cancer Imaging Archive (TCIA). [Online]. The cancer imaging archive. Source: https://public.cancerimagingarchive.net/ncia/login.jsf

Intelligence Planning for Aerobic Training Using a Genetic Algorithm

Nattapon Kumyaito[2] and Kreangsak Tamee[1,2(✉)]

[1] Research Center for Academic Excellence in Nonlinear Analysis and
Optimization, Faculty of Science, Naresuan University, Phitsanulok, Thailand
kreangsakt@nu.ac.th

[2] Department of Computer Science and Information Technology, Faculty of
Science, Naresuan University, Phitsanulok, Thailand
Thailandnattaponk56@email.nu.ac.th

Abstract. A training plan is an important part of aerobic training. A training plan with a good sequence of high intensive training sessions and low intensive training sessions will substantially raise athletic performance. A creation of training plan require a sport scientist or a sports coach to do. An athlete who trains with limit in sports science knowledge may get injury. In this study, we propose a systemic implementation using a genetic algorithm (GA) to find optimal training plan. Comparison of this study result and an independently created, apparently reliable training plan, it reveal that GA is obtain capability to find optimal training plan.

Keywords: Genetic algorithm · Training plan · Sport training · Endurance training · Aerobic training · Athlete performance modeling · TRIMP · Fitness · Fatigue

1 Introduction

A training plan is an important part of aerobic training. A training plan can be considered as a schedule of training sessions over a period of time. The information about each training session includes training intensity, duration and frequency, which are generally known as the training dose. The training dose information is used to quantify the training load for each training session. Professional athletes usually monitor their training data to assure themselves that they do not train more or less than stated in the training plan. Without this monitoring activity, they will have more risk to become overtrained or undertrained which may delay their preparation to an important race.

A good training plan normally adopts an athletics performance model [1, 2]. These models describe the relationship between the athlete's performance and their fitness and fatigue, which are the product of the training load. The athlete's performance value is equal to their chronic fitness value minus by their chronic fatigue value. Since the decay rate of fitness and fatigue is different, professional athletes must train with a good sequence of hard work followed by sufficient rest. This ensures a high chronic fitness value while minimizing the chronic fatigue value.

© Springer International Publishing AG 2018
T. Theeramunkong et al. (eds.), *Advances in Natural Language Processing,*
Intelligent Informatics and Smart Technology, Advances in Intelligent Systems

Although endurance sports demand extremely high aerobic fitness, athletic performance gradually increases over time. This phenomena requires a long term aerobic training plan. Normally, intermediate level athletes train for 8–12 weeks. To raise their athletic performance requires a training plan well in excess of this period.

Creating a sophisticated training plan demands knowledge of the sports science of the particular sport. In this case, the training plan creator needs knowledge of the sports science for the particular endurance sport, such as running, cycling and swimming. A professional cyclist training plan would normally be created by professional cycling coaches or sport scientists. Amateur athletes, however, have limited access to professional coaches or sport scientists. These athletes might access a training plan from an apparently reliable resource on the Internet, but training without the proper sports science knowledge in mind might lead to injury.

A genetic algorithm (GA) may solve this problem by intelligently creating a training plan for amateur athlete. A GA is a population-based heuristic search algorithm using techniques inspired by natural evolution, such as selection, crossover and mutation [3]. GA's have been successfully applied to scheduling problems in sports [4, 5] and other domains [6–9].

This study presents a GA for creating a sports training plan to support amateur athletes who are able to have limited consultation with a professional coach or sports science specialist.

This paper has the following sequence; Sect. 2 presents the state-of-the-art of computational intelligence on sport training scheduling. Section 3 describes how to formulate sports training data as input to the GA. Section 4 presents the experimental approach and the result of the study, with Sect. 5 discussing the findings and conclusion.

2 Related Works

GA is algorithm for metaheuristic search or optimization by mean of simulated evolution. GA finds optimal solution by repeating processes of evaluation and manipulation on randomized population. Population get manipulated by processes that mimic nature evolution such as selection, crossover and mutation.

GA has been applied on many sports related scheduling problem. In a referee assignment problem, a sports tournament needs a system that can assigns the referees to the match based on many factors. Meng et al. [4] use GA to optimally assign referees to their preferred time in a volleyball tournament schedule. Their study result shows that the proposed algorithm can produces a good solution. Atan and Hüseyinoğlu [5] address a scheduling problem of football game and referees simultaneously by using a mixed integer linear program formulation to games by concerning specific rules in the Turkish league. Then Atan applies GA method by concerning about referees-related workload constraints. Their study result shows a good performance in term of computation time and objective function value.

GA also applied on many other domains. Humyun et al. [7] address permutation flow shop scheduling problem by propose genetic algorithm (GA) based Memetic Algorithm (MA). Their study result shows that the proposed algorithm can obtain better result. Mohtashami [8] proposed dynamic GA method for vehicle scheduling in cross docking systems. Many independent algorithm was proposed for each sub-process in a cross docking system. Their result reveal that their proposed algorithm performs better.

3 Planning Aerobic Training Problem

An aerobic training plan can be considered as a task schedule of sport training. A good training plan that enables athletes to achieve higher athletic performance normally contains a sequence of training sessions and recovery sessions. The sequence of training sessions in each week are mixed with recovery sessions to avoid the risk of overtraining. The key point of organizing the sequence of training sessions is increasing the training load gradually. The training load can be quantified by Banister's TRIMP (TRaining IMPulse) [10]. TRIMP uses an average heart rate and a duration of a particular training session as inputs to calculate the training load in arbitrary units. The TRIMP of each training session is used to calculate the athlete's performance according to the Banister athlete performance model.

3.1 Problem Formulation

The problem of planning aerobic training for the optimization of athletic performance is the subject of this paper which proposes a process of problem formulation for the GA technique.

3.1.1 Specifying the Problem

In this section, the simulated data for the training plan is defined. The defined training plan covers a period of 8 weeks of activity. A training session occurs once each day. The training session data includes average heart rate (HR) in units of beats per minute (bpm) and duration (D) of the activity in minutes (min).

For creating a training plan that customized to a particular athlete, HR need to be normalized by restricted to current fitness level of particular athlete. One of the most widely adopted is Andrew Coggan training zone [11]. This training zone is a proportion of functional threshold HR (FTHR) that ranged between resting HR and maximum HR. The resting HR collected immediately when an athlete wakes up while maximum HR and FTHR are collected from field test. Maximum HR is collected from a log of recently racing event or from a test that gradually increasing intensity until an athlete cannot progress anymore. FTHR is an average HR of highest intensity workout which an athlete can sustain for 30–60 min. Maximum HR is seem to be decreased when an athlete getting older while resting HR and FTHR are adaptable to training progress. Professional athlete normally have lower resting HR and higher FTHR. Thus, for the same training zone, athletes might have different HR value. Each training zone have demonstrate in Table 1. This experiment adapts Coggan's training zone as a method to normalize HR value that personalize to a particular athlete.

Table 1. Andrew Coggan training zone

Intensity	Average HR	Description
Level 1	<68%	Active recuperation
Level 2	69–83%	Endurance
Level 3	84–94%	Tempo
Level 4	95–105%	Lactate threshold
Level 5	>106%	Maximal aerobic power

In this experiment, athlete's exercise data was simulated for an athlete who has resting HR at 45 bpm, maximum HR at 192 bpm and FTHR at 174 bpm. The training duration range was between 0–300 min.

The range of HR was divided into 10 zones which were modified from Andrew Coggan's Training Zones by divided each zone in a half as showed in Table 2. The range of D was also divided into 10 zones as showed in Table 3.

Table 2. Heart rate zone

HR Zone	HR (bpm)	HR (% of FTHR)
0	45–82	25.86–47.13
1	83–118	47.70–67.82
2	119–132	68.39–75.86
3	133–144	75.86–82.76
4	145–155	83.33–89.08
5	156–164	89.66–94.25
6	165–174	94.83–100.00
7	175–183	100.57–105.17
8	184–188	105.75–108.05
9	189–192	108.62–110.34

Table 3. Duration zone

D Zone	D (min)
0	30
1	60
2	90
3	120
4	150
5	180
6	210
7	240
8	270
9	300

3.1.2 Chromosome Encoding

In this experiment, a training plan was encoded to a chromosome. 300 training plans or 300 chromosomes, were randomly initiated as initial population. Each chromosome consisted of 56 training sessions, and each training session indicated a target of related exercise values which an athlete should do in daily training. Related exercise values in each training session were a pair of genes that represent an average HR zone value and a D zone value. Each training plan was encoded into a chromosome with 112 genes. Each training session was represented by 2 consequence genes as demonstrate in Fig. 1.

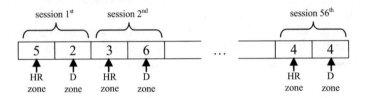

Fig. 1. Example of encoding aerobic training plan to chromosome

By this encoding method, the problem domain data encoded into a chromosome. Target workout values for each training session were normalized and transform to a pair of genes that was suitable for processing in GA operations.

3.1.3 Fitness Function

The fitness function in this experiment used Banister's model of elite athletic performance [1] to evaluate individual chromosomes. This model uses TRIMP to quantify the training load of each training session. The TRIMP equation is described in Eq. 1.

$$w_i = D \times HR_r \times 0.64 e^{y \times HR_r} \tag{1}$$

TRIMP (w_i) is an arbitrary value representing the training load of the training session i. It is the multiplication product of duration (D), average heart rate (HR_r), and a model constant (y). HR_r is the normalized heart rate in the range between resting state and maximum heart rate. Constant y is a HR_r multiplier which is equal to 1.92 for men and 1.67 for women. These constants were developed based on the experimentally observed relationship between heart rate and lactate level. For men this will give a TRIMP value of 0–4.37 per minute and for women 0–3.4.

After getting the TRIMP value of each training session, athlete performance is evaluated by the Banister model according to the following equation.

$$p_t = p_0 + k_1 w_0 e^{-t/r_1} + k_1 \sum_{i=1}^{t-1} w_i e^{-(t-i)/r_1} - k_2 \sum_{i=1}^{t-1} w_i e^{-(t-i)/r_2} \tag{2}$$

An athletic performance (p_t) is a summation of an athlete's basic level performance (p_0), past training load (w_0), training load (w_i) that is stimulated from training for *t* sessions according to the training plan. Training load had influence performance by 2 mean, rise up performance by improving fitness and drag down performance by suffering fatigue. This antagonist model represent these influences of training load by positive (fitness) terms and negative (fatigue) terms. k_1 and k_2 represent the coefficient of fitness and fatigue subsequently. r_1 and r_2 are the decay rates of fitness and fatigue subsequently. In this study, all model parameters are based on result of model fitting from Busso et al. [2].

The fitness value of each chromosome was evaluated by these following step.

1. Decode each genes according to their represented value. If that gene is represented HR zone value, it decode to its zone upper bound value. For example, if HR genes is 5, it decode to 164 bpm for HR value. For duration genes, it simply decoded to its duration value. For example, if D zone is 8, it decode to 270 min.
2. Evaluate TRIMP value of each training session by evaluating each pair of consequence genes, HR and D genes, as a training session. Iterate throughout every pairs of genes in a chromosome for a list of TRIMP values of each training sessions. In this study, the result of this step is a list that contain 56 TRIMP values.
3. Evaluate performance of a chromosome by using a result from previous step. TRIMP value of each training session were used as a training load for a particular training session in Eq. 2.

3.1.4 GA Operation

At the beginning of the GA computations, each chromosome in the initial population was randomly created as an encoded chromosome as defined in Sect. 3.1.2. The initial population was processed through several GA operations as in the following pseudo code.

```
evaluate(population)
for i = 0 to 300{
    population = selectTournament(population
                , offspringSize=len(population)
                , tournamentSize=3)
    offspring = crossover(population
                , cxOnePoint
                , cxProbability=0.65)
    offspring = mutation(population
                , mutUniformInt
                , mutProbability=0.1)
    evaluate(offspring)
    population = offspring
}
```

First, the initial population was evaluated by the fitness function as defined in Sect. 3.1.3. Second, 3 chromosomes in the sorted population were randomly selected in a tournament. The best fitness chromosome in tournament is selected to be added to the offspring. This selection process continue until offspring size is equal to the initial population size. Third, the offspring was evolved by a two-step evolution. (1) Mated pairs of consecutive chromosomes by a one point crossover operation at a 65% probability. Resulting children replaced their respective parents. (2) Mutate the chromosomes by a random value that ranged between 0 and 9 at 1% probability. Fourth, evaluate all offspring from the previous step by using the fitness function. Finally, the final offspring is used as a population for the next loop of the evolution.

4 Experiment and Result

In this experiment, Python script was implemented by using the Distributed Evolutionary Algorithms (DEAP) framework version 1.0.2 [12]. Python was implemented and tested on WinPython 64 bit version 3.4.3.5. Graphs were created by the Pyplot library, included in Matplotlib library [13].

A population of 300 chromosomes were randomly created as a string representing the chromosomes. The string—chromosome encoding was described in Sect. 3.1.2. The population was evolved for 300 generations by GA as pseudo code in 3.1.4. The final result of GA is demonstrated in Fig. 2. This graph present statistics of chromosome fitness value include maximum, minimum, average, and standard deviation.

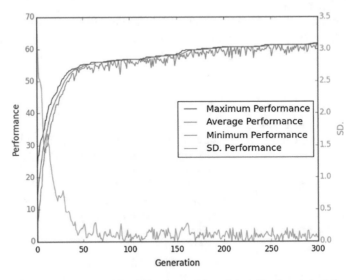

Fig. 2. Athletic performance of intelligence aerobic training plan by each GA generations

The GA began to be unable to find a better aerobic training plan at about generation 75, where the athletic performance began to struggle to show significant improvement. Termination of the evolution was considered at generation 75, after which athletic performance showed no significant improvement.

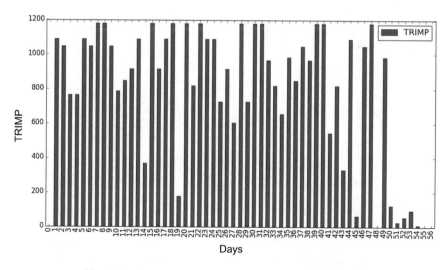

Fig. 3. TRIMP score in each day of the best training plan

The TRIMP of each training session in the best aerobic training plan that GA created is shown in Fig. 3. The sequence of the TRIMP values reflects the training load of each training session. The result shows a good sequence of intensive training sessions followed by lower intense training. This may imply that GA is capable of creating a good training plan that taking overcompensation into account. The result appears a good sequence of exercises followed by sufficient rest. This sequence of training sessions substantially raises athletic performance to the last training session of the aerobic training plan as demonstrated in Fig. 4.

The experiment on various population sizes was executed to investigate the behaviors of GA on chromosome-encoded training plan. This experiment sets the size of offspring to be equal to the size of population for 300 generations. The comparison of various population in Fig. 5 shows that higher population size make GA find its optimal solution faster and its solutions have higher fitness value.

5   Discussion

The result of this experiment shows that GA has the capabilities to create a good aerobic training plan. The aerobic training plan created by GA had a good sequence of high intensive training sessions followed by low intensive training sessions. As a result of this experiment, it can be stated that athletes who train by following the training plan created by GA will raise their athletic performance to a much higher level.

Comparing the GA training plan with an independently created, apparently reliable training plan, such as the British Cycling training plan for intermediate and advanced cyclists [14], Both of them share the pattern of a sequence of high intensive training and low intensive training. The British Cycling training plan also has some intensive training sessions which followed by lower intensive training sessions or a rest day (see

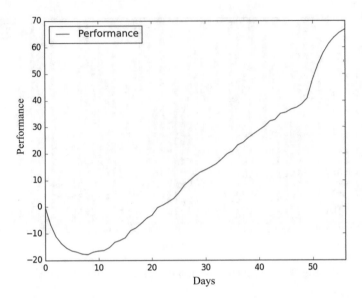

Fig. 4. Athletic performance progression of the best training plan

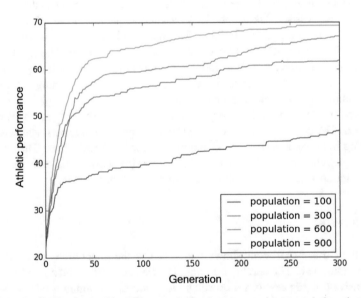

Fig. 5. Comparison of best fitness solution between various population sizes

Fig. 6). This training pattern substantially raises athletic performance at the last training session of the aerobic training plan. The comparison between GA training plan and UK training plan has presented in Table 4.

One potential improvement opportunities for the GA approach is adopting some constraint-handling techniques. The GA training plan might need to be restricted to human physical abilities. Created training plan need to be customized for an individual athlete if the aerobic training plan is found to be too hard for the athlete to follow. For example, in the 1st training session of the training plan as shown in Fig. 3, training at 87.41% of maximum HR for 5 h is specified. However, training at this level of intensity is very hard to accomplish by many athletes not yet at the highest levels of performance.

The direction of future work on the intelligent planning for aerobic training by GA can be the creation of good training sequences that raise high athletic performance while showing concern for actual human physical abilities. GA and other nature-inspired meta- heuristic were designed to dealing with unconstraint search spaces [15, 16]. These algorithm focus on only optimized solutions. So that, some constraint restricted solution might be overlooked. Some constraint-handling techniques can cause the search method to keep away focus on only optimal solutions that concern human physical abilities to just feasible solutions that too hard for human. This techniques might make GA to be able to considering on feasible solutions instead optimal solutions.

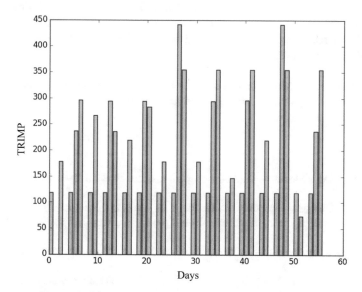

Fig. 6. TRIMP score in each day of UK training plan (use minimum value of HR range and D range for calculate TRIMP score)

By this direction, the training sessions in the aerobic training plan might be a guide to more effective training methods for athletes.

Table 4. Comparison of TRIMP value between GA training plan and UK training plan

Training plan	TRIMP				Fitness	Fatigue	Performance
	Max	Min	Mean	S.D.			
GA	1178.662	0	686.174	359.670	22,979.291	1203.004	47.584
UK	441.265	0	151.273	128.948	5538.294	999.152	5.653

6   Conclusion

An aerobic training plan is a set of training guidelines which athletes use to improve athletic performance. According to athletic performance models and overcompensation effects, training plans usually consist of a good sequence of intensive exercises and sufficient rests. A training plan can be considered as a scheduling problem. This problem has opportunities for GA to find optimal training plans that significantly raise athletic performance.

In this experiment, GA showed some signs of being capable to schedule a training plan that can result in a high performance athlete.

This is a very first step of the GA approach for intelligently creating a training plan. The results of this experiment also reveal some research opportunities that more advanced GA techniques can be adopted to find better training plans that consider human physical abilities.

Acknowledgements. Many thanks to Mr. Roy Morien of the Naresuan University Language Centre for his editing assistance and advice on English expression in this document.

References

1. Banister, E.W.: Modeling elite athletic performance. Physiol. Test. Elite Athl. 403–424 (1991)
2. Busso, T., Denis, C., Bonnefoy, R., Geyssant, A., Lacour, J.-R.: Modeling of adaptations to physical training by using a recursive least squares algorithm. J. Appl. Physiol. **82**, 1685–1693 (1997)
3. Goldberg, D., Holland, J.: Genetic algorithms and machine learning. Mach. Learn. **3**, 95–99 (1988)
4. Meng, F.-W., Chen, K.-C., Lin, K.-C., Chen, R.-C.: Scheduling volleyball games using linear programming and genetic algorithm. Inf. Technol. J. **13**, 2411 (2014)
5. Atan, T., Hüseyinoğlu, O.P.: Simultaneous scheduling of football games and referees using Turkish league data. Int. Trans. Oper. Res. (2015)
6. Zhang, R., Ong, S.K., Nee, A.Y.C.: A simulation-based genetic algorithm approach for remanufacturing process planning and scheduling. Appl. Soft Comput. **37**, 521–532 (2015)
7. Rahman, H.F., Sarker, R., Essam, D.: A genetic algorithm for permutation flow shop scheduling under make to stock production system. Comput. Ind. Eng. **90**, 12–24 (2015)
8. Mohtashami, A.: A novel dynamic genetic algorithm-based method for vehicle scheduling in cross docking systems with frequent unloading operation. Comput. Ind. Eng. **90**, 221–240 (2015)

9. Pattanayak, P., Kumar, P.: A computationally efficient genetic algorithm for MIMO broadcast scheduling. Appl. Soft Comput. **37**, 545–553 (2015)
10. Ew, B., Tw, C.: Planning for future performance: implications for long term training. Can. J. Appl. Sport Sci. **5**, 170–176 (1980)
11. Allen, H., Coggan, A.: Training and Racing with a Power Meter. VeloPress (2010)
12. Fortin, F.-A., Rainville, D., Gardner, M.-A.G., Parizeau, M., Gagné, C.: DEAP: Evolutionary algorithms made easy. J. Mach. Learn. Res. **13**, 2171–2175 (2012)
13. Droettboom, M., Hunter, J., Firing, E., Caswell, T.A., Elson, P., Dale, D., Lee, J.-J., McDougall, D., Root, B., Straw, A., Seppänen, J.K., Nielsen, J.H., May, R., Varoquaux, G., Yu, T.S., Moad, C., Gohlke, C., Würtz, P., Hisch, T., Silvester, S., Ivanov, P., Whitaker, J., Cimarron, W., Hobson, P., Giuca, M., Thomas, I., Mmetz-bn, E.J., Evans, J., Hyams, D., Nemec, N.: Matplotlib: v1.4.3. (2015)
14. Introduction to the Foundation Plan for Intermediate/Advanced Riders. https://www.britishcycling.org.uk/knowledge/article/izn20140929-Training-Introduction-to-the-Foundation-Plan-for-Intermediate—Advanced-0
15. Bäck, T.: Evolutionary Algorithms in Theory and Practice: Evolution Strategies, Evolutionary Programming. Genetic Algorithms. Oxford University Press, Oxford, UK (1996)
16. Engelbrecht, A.P.: Fundamentals of Computational Swarm Intelligence. Wiley (2005)

Palm's Lines Detection and Automatic Palmistry Prediction System

Tanasanee Phienthrakul[✉]

Department of Computer Engineering, Faculty of Engineering, Mahidol
University, Salaya, Nakorn Pathom, Thailand
tanasanee.phi@mahidol.ac.th

Abstract. This paper presents a method for palm and palm's lines detection
based on image processing techniques. An application of the proposed method is
illustrated in the automatic palmistry system. Both hardware and software are
created and tested. The system can detect palm and three main lines, i.e., life
line, heart line, and brain line. Line's position, line's length, and line's curvature
are used for palmistry prediction. These three lines will be compared to the lines
in the line pattern archives by using the nearest neighbor method. The experimental results show that this system can detect palm and palm's lines and this
system yields the suitable results on many examples. Furthermore, the concept
of this system can be applied to identification and authentication in security
approaches or in the embedded system fields.

Keywords: Palmistry · Palm detection · Line detection · Image processing ·
Nearest neighbor

1 Introduction

Palmistry is the art of foretelling the future through the study of the palm lines [1], also
known as palm reading or chirology. This practice is found all over the world with the
various cultures. The objective is to evaluate a person's character or future by studying
the palm. Someone believes in this foretelling and a lot of people are funny in them.
There are many ways to gain insight from the hand, such as, talking with the foretellers,
reading from the books, and using the mobile applications. These ways may give the
different forecasts depend on the experience of foretellers.

In general, there are many lines that can be seen on a palm. The most important and
popular of all the lines are the heart line, the brain line, and the life line as shown in
Fig. 1. Heart line can be read in either direction depending on the tradition being
followed. This line is believed to indicate emotional stability, romantic perspectives,
depression, and cardiac health [2]. Brain line represents a person's learning style,
communication approach, intellectualism, and thirst for knowledge. A curved line is
associated with creativity and spontaneity, while a straight line is linked with practi-
cality and a structured approach [2]. Life line begins near the thumb and travels in an
arc towards the wrist. This line reflects physical health, general well-being, and major
life changes [2]. However, its length is not associated with length of life [2].

© Springer International Publishing AG 2018
T. Theeramunkong et al. (eds.), *Advances in Natural Language Processing,
Intelligent Informatics and Smart Technology*, Advances in Intelligent Systems

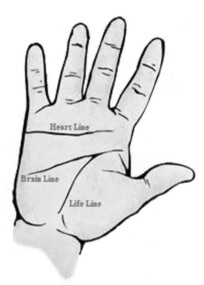

Fig. 1. Three Main Lines in a Hand

Although there are some applications for palm reading such as Palm Reading PRO [3], Palmist True Vision [4], The Japanese Palmistry [5], and Palmistry Fortune Predict [6], those applications are created for entertainment. The systems that are already available based on palmistry are either not automated or not accurate. Moreover, the correctness of those applications was affected by the skill of user and the performance of camera.

For hand detection, there are some researchers that try to develop hand detection techniques. In a research of Dhawan and Honrao [7], they presented the methods for human computer interaction using the user's hand. Hand was segmented by Otsu Thresholding. Contour and convex hull techniques were used to detect the hand. Then the gestures are detected. The gestures can be used to control the robot. This research showed that the image processing can be used to detect the hand and it can be applied to real application. Furthermore, the hand detection can be used for multiple purposes such as determining human layout and actions, human temporal analysis, and recognizing sign language.

This paper proposes a prototype for the palmistry machine and develops the palmistry software that used some image processing techniques. This system can give the foretelling results from a hand. The used techniques and experimental results are described in this paper.

2   Automatic Palmistry Prediction System

The automatic palmistry prediction system in this paper composes of 2 main parts, which are palmistry machine and palmistry system. Figure 2 showed the design of our palmistry system and the components of the palmistry machine.

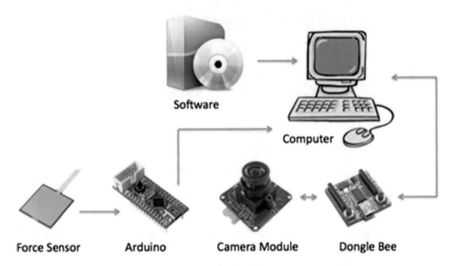

Fig. 2. Components of palmistry system

2.1   Palmistry Machine

For the palmistry machine, the force sensor will be used for the hand existing detection. When the users place their hand, the camera will capture their hand image. Dongle Bee is used for signal transformation. Dongle Bee can be used for programming or communicating with the applications. From the first idea, we plan to use Arduino for sending the image but we found that it took a long time for waiting (about 4 min per image). Hence, the Dongle Bee is applied and it takes about 20 s for sending an image from our machine to the computer. Then, the image will be processed by our software and show the results on the monitor.

Within the palmistry machine, there is a camera module that will capture the image when the command is sent to it. There are five LEDs to increase the brightness of the captured image. Both camera and LEDs are settled in the different positions. In the first experiment, the LEDs are installed on 4 sides of the box and the camera is at the bottom of the box, as showed in Fig. 3. The capture image from this setting is showed in Fig. 4. There are some spots of light that may disturb the image processing in the next steps.

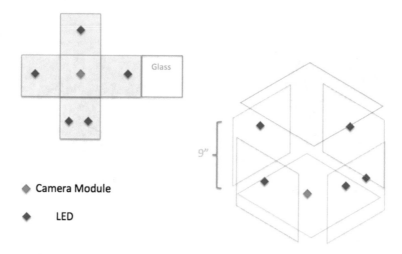

Camera Module

LED

Fig. 3. The first design of palmistry machine

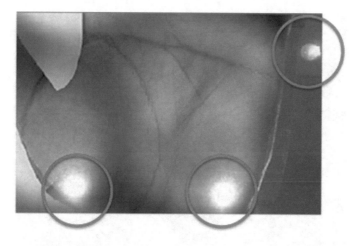

Fig. 4. Example of the captured image from the first design

For the second design, the LEDs are moved to a side of the box and the camera is moved up from the bottom as showed in Fig. 5. With this design, the spots of light are not affected to the captured image. They are on the top of image that is the area of fingers and they are not used for palmistry prediction. The palm's lines are clear and they can be used to process in the next steps. An example of the captured image is showed in Fig. 6.

However, we notice that if the users place their hand on the unsuitable position, the palm area in the captured image may not complete. Thus, we decided to move the camera back to the bottom of the box. The insignificant area that may appear on the captured image will be removed by image processing techniques in the later. Hence, the

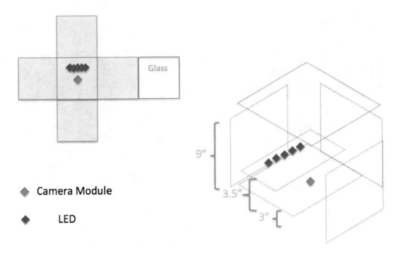

Fig. 5. The second design of palmistry machine

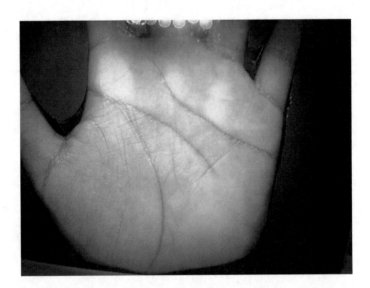

Fig. 6. Example of the captured image from the second design

final design of the palmistry machine will have the glass at the top of the machine. The camera module was installed at the bottom of the box. There are five LEDs in the box as showed in Fig. 7. The heights of box and LEDs depend on the focus' length of the camera; for this research, they are 9 and 3.5 inches, respectively. An example of the captured image from the last design is showed in Fig. 8.

To control the light from the outside, we use the black cover as showed in Fig. 9. Users must place their hand on the glass within the black cover. Then, they must wait until the LED change from red color to be green color as showed in Fig. 10. When the

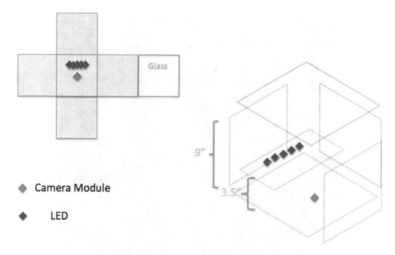

Fig. 7. Design of palmistry machine

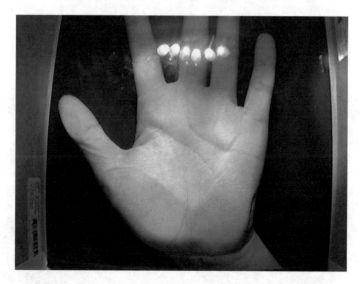

Fig. 8. Example of the captured image from the last design

LED is green, the image will be sent to the computer and the palmistry software will be processed.

2.2 Palmistry System

To give the prediction results from the palm image, we propose 7 steps of image processing. These steps are used for the different purposes such as for palm detection, for line detection, or for pattern matching. These steps are described in the following.

Fig. 9. Palmistry machine

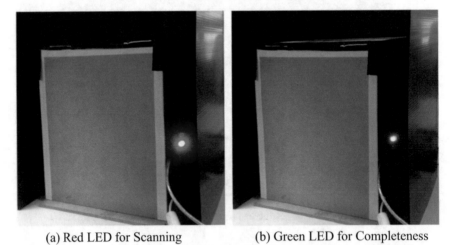

(a) Red LED for Scanning   (b) Green LED for Completeness

Fig. 10. LED for working status

1. The image from the camera is saved as .jpg file. An example of palm image is showed in Fig. 11.
2. Haar-like is used to find the position of palm in the image. Then the image is cropped; only the palm is recorded as showed in Figs. 12 and 13. The palm image is changed to gray scale image to reduce the processing time.

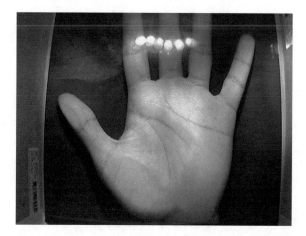

Fig. 11. Example of palm image

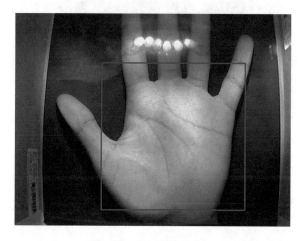

Fig. 12. Haar-like results

3. Noises are removed by using scale down and scale up techniques. First, the image will be scaled down. Thus the small size of pixels will be removed. Then, the image will be enlarged to the original size and the noises are removed. A result is showed in Fig. 14.
4. Canny edge detection is used to find the line in the image as showed in Fig. 15.
5. Dilation and erosion are used to enlarge the thickness of palm line. The lines are easier to find as showed in Fig. 16.
6. Grid lines are used to divide the image into 10×10 parts as showed in Fig. 17. These grid cells are used to find the positions and the contours of three lines, i.e., life, heart, and brain lines as showed in Fig. 18.
7. The position and the curvature of lines are compared to lines in the line pattern archives. Each line is detected by template matching. Examples of line patterns in

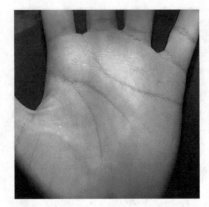

Fig. 13. Cropped image

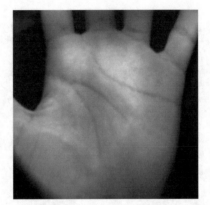

Fig. 14. Gray scale image

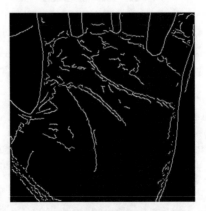

Fig. 15. Canny edge detection

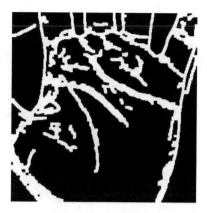

Fig. 16. Dilation and erosion

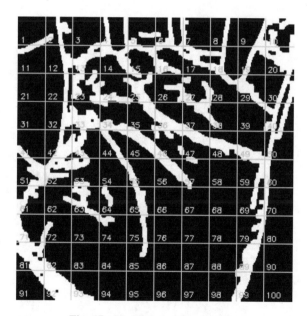

Fig. 17. Divide into 10 × 10 parts

archives are showed in Fig. 19. The nearest pattern will be used to find the prediction results. The examples of palmistry prediction outputs are showed in Figs. 20, 21 and 22.

Fig. 18. Find 3 Palm's lines

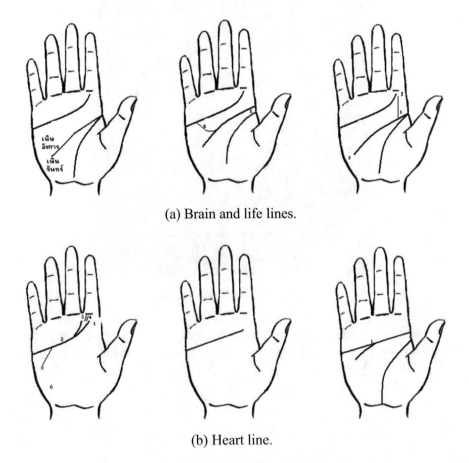

(a) Brain and life lines.

(b) Heart line.

Fig. 19. Example of line pattern archives

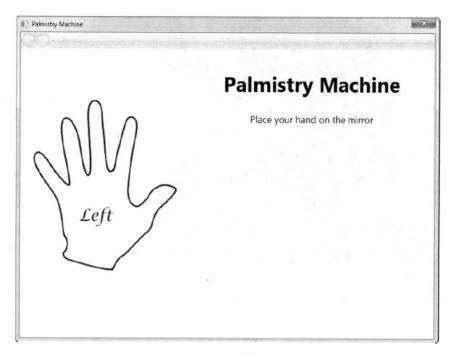

Fig. 20. Palmistry software start page

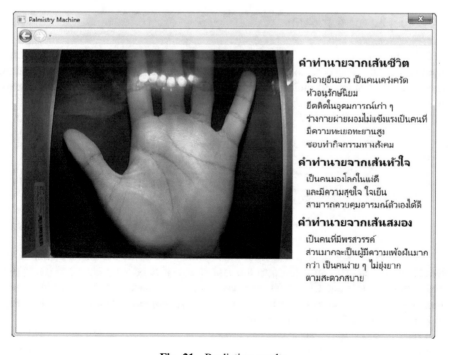

Fig. 21. Prediction results

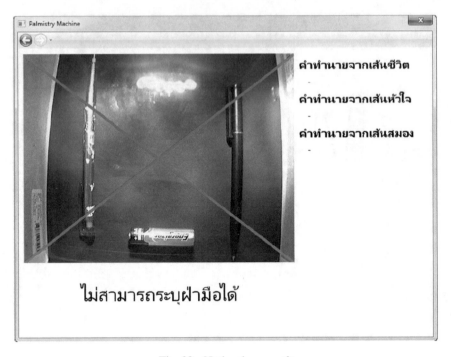

Fig. 22. No hand, no result

3   Experimental Results

In order to confirm that the proposed method is suitable for using in the palmistry prediction system, we conduct some experiments. The 160 images are used for testing. There are 82 palm images and 78 other images. The results of palm detection are shown in Table 1.

Table 1. Confusion table for palm detection

		Actual	
		Hand	Not hand
Predict	Hand	80	13
	Not hand	2	65

The experimental results show that the total accuracy is 90.625%. Some hands cannot be detected because the position and shape of hand may not be suitable. For other objects, there are 13 images that our method was deceived. These objects have the similar shapes. Some objects have the texture that similar to palm's lines. This may be improved by adding the training data.

Then, to evaluate the performance of line detection, we compared the results of line detection from our system with our observation. The numbers of correction line were recorded and showed in Table 2.

Table 2. Line detection results

	Life line	Heart line	Brain line
Correct	111	127	97
Incorrect	49	33	63
Percentage accuracy	69.375	79.375	60.625

The average accuracy of palm's line detection is 69.79%. Some mistakes come from the position of line and the quality of images. Sometimes, we cannot detect the lines because the lines are very thin and short. Many hands have a lot of small lines that make the confusion for line detection.

4 Conclusion and Future Work

An automatic palmistry prediction system was proposed. This system composed of both hardware and software. The users can place their hands and get the prediction results. However, the actual objective is to propose a guideline for palm detection and palm's line detection. Haar training is used for palm detection the experimental result show that palm area can be detected with the high accuracy. Then, the positions of lines in hand are searched. The concept of zone analysis is applied. Each line is compared with the line pattern archives; the nearest pattern is selected and the prediction results are showed. Although the application of this method is designed for fun, this method can be applied for the security purposes. This method can be a part of identification and authentication or it can be applied to the investigation tasks. However, the average accuracies of palm's line detection are not high for the security system. The palm line detection should be improved for higher accuracies.

Acknowledgements. I would like to thank to Kittitat Kupta-arpakul and Thanatarn Chuai-songkro for hardware preparation and data collection. Both of them are students at department of computer engineering, faculty of engineering, Mahidol University, and graduated in education year 2014 and 2013, respectively. This work may not success if I lack their dedication.

References

1. Navpat, A.K., Mukherjee, R., Pandita, V., Gupta, S.: Application of prediction software in palmistry. In: MPGI National Multi Conference 2012 (MPGINMC-2012), Proceedings published by International Journal of Computer Applications (IJCA), pp. 6–8 (2012)
2. How to Read Palms. http://www.wikihow.com/Read-Palms. Accessed 18 June 2014
3. Palm Reading PRO. https://play.google.com/store/apps/details?id=hydrasoftappstudio. palmreading. Accessed 16 June 2014

4. Palmist True Vision Pro 2. https://play.google.com/store/apps/details?id=com.ttn.palmistry. Accessed 16 June 2014
5. The Japanese Palmistry. https://play.google.com/store/apps/details?id=jp.co.unbalance.tesou. Accessed 16 June 2014
6. Palmistry Fortune Predict. https://play.google.com/store/apps/details?id=com.Palmistry_Fortune_Predict_207094&hl=th. Accessed 16 June 2014
7. Dhawan, A., Honrao, V.: Implementation of hand detection based techniques for human computer interaction. Int. J. Comput. Appl. (0975–8887), **72**(17), 6–13 (2013)

The Algorithm for Financial Transactions on Smartphones Using Two-Factor Authentication Based on Passwords and Face Recognition

Mahasak Ketcham and Nutsuda Fagfae[✉]

Department of Information Technology Management, Faculty of Information Technology, King Mongkut's University of Technology North Bangkok, Bangkok, Thailand
mahasak.k@it.kmutnb.ac.th, nutsuda.mint@gmail.com

Abstract. This research applied the two-factor authentication with rather high security to mobile phones. The approach was the combination of face recognition and passwords converted to byte of array before it changed to string. Then, it was stored in the database for the security and storage convenience. Regarding the authentication through face recognition, Eigenface's technique was employed here to develop the system for its higher accuracy. The results revealed that the program could recognize face accurately with the mean at 98%, considered as a very high accuracy and at an excellent criteria. The whole processing time was 5.5432 s, which was assumed as the medium speed and agreeable criteria.

Keywords: Face recognition · Authentication · Smartphone · Two-factor authentication · Eigenface · Byte of array

1 Introduction

Smartphones have been regarded as indispensable communication devices for people's lives, particularly for sending information and news. Above all, they are basically designed with multifunctions equivalent to computers. Search, commucations, transactions, or even educations, for instance. People mostly access to applications in smartphones by entering their usernames and passwords. Such authentication might not be adequate for vital systems such as online banking due to disadvantages of passwords. To clarify, they can simply be hacked without the awareness of owners. That is why other components are needed to reinforce the security. Biometrics can be used to examine physical appearances or human behaviors in order to confirm identities with high security. At present, there are diverse biometric systems, e.g., fingerprints, faces, face, etc. Focusing on face, it is manipulated for identity confirmation as it is unique and difficult to be copied. For this reason, the researcher applied face recognition as well as passwords to the two-factor authentication for the solution of electronic information security hacking.

According to other relevant researches, there are 3 groups of authentication system development, namely, using passwords and using face recognition. The details are as follows:

Group 1 [1–3]: The technique developed by Neural networks In classification specifically by creating a tool that is capable of learning, pattern recognition and the creation of new knowledge. But the results may not be as good as it should. If the image size is too small. And not be available to analyze face images that are more than one face.

Group 2 [4–10]: Use Eigenface's technique, Eigenface a technique to facial recognition, which is another method that has been popular because is a simple but effective recognition as well. The principle function of PCA.

Group 3 [11–13]: It is the method of entering password, to which cryptography is performed in order to protect information or default messages to be sent to receivers without others 'comprehension. The principles of symmetric key algorithms are brought to help it work faster and become easier to be operated. Similarly, it is also a one-factor authentication.

So, the researcher came up with the idea to initiate a two-factor authentication on smartphones for double identity confirmation, and to efficiently reduce the problem of hacking. Eigenface's technique was conducted here because face stripes must be examined with high accuracy and least errors. Symmetric key algorithms was also applied in the research for the security and storage convenience.

2 Authentication

Authentication is the primary step of security control during the procedure. It basically takes users' evidences to verify who those persons or those asserters are, and whether they are permitted to access resources in the system or not. There are many types of authentication, for example, authentication based on passwords, authentication based on individuals' biometric identifications or one-time passwords, for example. Each type contains dissimilar advantages and disadvantages. It is all up to the necessity of usage. For an open network like the Internet, authentication is regarded as the most important initial process to protect network security. It holds special properties for the system confirmation. User Authentication It is to prove that users are truly permitted to access the system. The mechanisms of authentication are described as per below:

Possession factors = a category of credentials based on items that the user has with them, typically a hardware device such as a security token or a mobile phone used in conjunction with a software token.

Knowledge factors = a category of authentication credentials consisting of information that the user possesses, such as a personal identification number (PIN), a user name, a password or the answer to a secret question.

Biometric Factor = It is the method of using the different biometric identification of each person to confirm the access to the system. Manipulating biometric identifications for this examination is called biometrics. Individual's information is scanned, and then compared with the one stored in the system. If the information of both sources is found related, it might be able to conclude that the authentication is highly accurate.

3 Proposed Method

3.1 The Perspectives of the System

The procedure of face recognition system was the design of the system's perspectives. The operation was separated into 2 parts sharing the same function in view of information verification. The difference was that the information of users was stored in the database. Users must identify their names as well as passwords when they used the system at the first time. Then, face scan process was taken, and the results would be converted to byte of array as the form of string. After that, they were stored in the database for the security and storage convenience. When the system was used, File Stream was stimulated to retrieve the information in the database stored as byte. It was subsequently compared with face stripes of the originals and of the database. If the values came out within the determined rates, the system will allow users' access (Fig. 1).

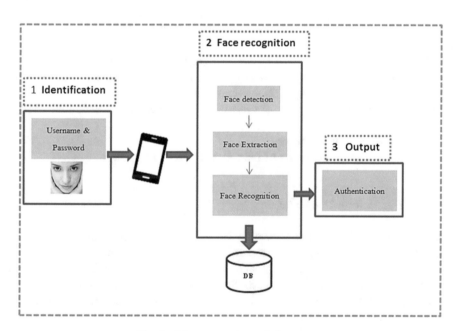

Fig. 1. The perspectives of the system

3.2 Below is the Authentication Procedure

3.2.1 Identity with a Username—Password of the Following Steps

As part of the identification with encryption. In this design, the researchers have developed a technique Symmetric Key Cryptography to encrypt the data with only the sender and receiver. By working from users has been put into the system, then the system will be decrypted using the technique of replacing the letter next to the first position, as if the letter A was changed to B, who agreed to take shape. a encrypted with a secret key of the sender and recipient (Fig. 2).

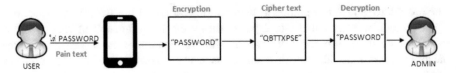

Fig. 2. The work on the part of the identification with encryption

In this research this research was designed to provide a user name can't be duplicated., the comparison between patterns of users in the test and those of users in the system was carried out. The evaluation was grounded on threshold values of the system as per Eq. 1 presented below:

$$User \begin{cases} User; if \min(Dist) \leq Threshold \\ reject; if \min(Dist) > Threshold \end{cases} \tag{1}$$

User here was resulted from the system as per the following explanation:

User It referred to user series i who possessed the lowest distance value, and the value was lower than or as equal as the threshold value of the system.

Dist: It was the distance value of user series 1 when compared with a tested user.

3.2.2   Face Recognition Process

Face recognition process consists of the following steps (Fig. 3).

3.2.2.1   Face Information Collection

The process of face information collection was to bring in images for image processing. The images of face employed in this research were from Smartphone camera 10 people, both men and women each of the images is not the same. It will face either keep a straight face, the left side, Right side, Smiley, page laughs etc. This is the tone for the processing, distribution frequency of the intensity and regularity (Histogram Equalization) to adjust the contrast of the image and enhance the contrast of the image before processing the next step.

3.2.2.2   Face Detection

As for the elements of the face such as the eyes, mouth, etc. The model has two kinds to find positions are as follows:

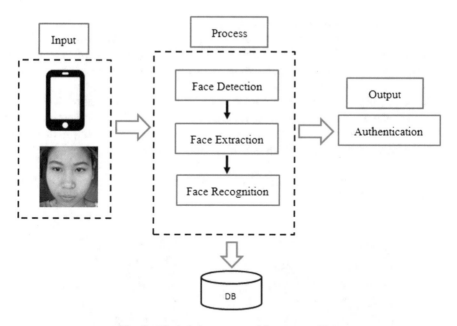

Fig. 3. The whole process of face recognition

3.2.2.2.1   *Geometry Model of an Eye-Mouth Triangle*

In order to calculate the position of the eyes and mouth are the most valuable was selected perfect Triangle. See Fig. 4.

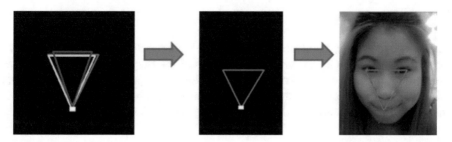

Fig. 4. Display locate eyes and mouth with triangular geometric modeling of the eyes and mouth

3.2.2.2.2   *Geometric Face Model*

Put face images to the processing geometry model. When the value of Geometry model of an eye-mouth triangle, and then calculates the position of the mouth. Nose by building relationships with the values obtained from the positions of both eyes following steps.

- Find the midpoint between the two eyes.
- Create a circle with a radius is half the distance between the two eyes. To the intersection on the Y-axis of the circle is the location of the nose, which is likely to be both the top and the bottom, but in this study using a straight face, so the bottom point only one point into consideration.
- Create an ellipse with a minor axis (the axis X) has a length equal to the distance between the eyes and the major axis (vertical axis Y) length is half the distance of both eyes, divide by 0.6. To the intersection on the Y axis of the ellipse defined as the position of the mouth, which is likely to be on the point and the point. Like the nose.
- Coordinates the several the central is calculated in a separate component of face images. See Fig. 5.

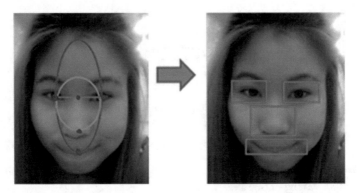

Fig. 5. Positions that are calculated by the geometry model

3.2.2.3   Face Extraction

The researcher split component of facial features is four section: the left eye, right eye, face, nose and mouth by showing different parts of the frequency, intensity, and then dividing the frequency range from 0 to 255 contiguous frequency range of 9 [10] have calculated the average of all 29 states and then bring together a collection of face images and other features. To learn test and recognition. See Fig. 6.

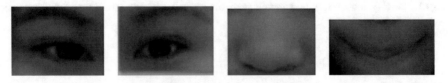

Fig. 6. The split component of the face

3.2.2.4   Face Recognition

In the process of face recognition. Use the Eigenface. Based on the work of the PCA, which were both two data sets were converted to a gray scale image, then make a calculation features of images to be used for recognition by the image the trainer P and image size M × N is x_1, x_2, ...,x_p and conversion matrix image tiles into a vector magnitude 1 × d, d = m × n use datasets trainer as $\mu = \{ \mu_1, \mu_2, ..., \mu_N \}$ then taken. calculate the average of the face image to determine the new data and then calculate the variance of the data matrix code of variance on the average, where C is the matrix of covariance μ and water bills. C to calculate the Eigen vector corresponding to the Eigen values correspond to Eigen descending. Then carefully remove the specific value that is not zero. Then remove the extracted features of images to be used for recognition of the two equations.

$$Y_k = U_d^T (\mu_k - \mu); k = 1, 2, ..., P \tag{2}$$

4   Experimental Result

The procedure of the development stated was tested composed of 100 volunteers. The system employed was advanced with Java. The test was done on Android smartphones, with FAR (False Accept Rate), FRR (False Reject Rate), EER (Equal Error Rate), and CRR (Correct Recognition Rate) as the instruments to measure system accuracy. The test was shown in Table 1.

Table 1. The results of face recognition efficiency

	Identification mode			
Face image	FAR (%)	FRR (%)	EER (%)	CRR (%)
Face image of 100 person	1.859162	2.316984	2.149726	98.635142

The results revealed that the two-factor authentication was compared with the traditional system requiring usernames and passwords. The processing time was brought as the indicator. See Table 2.

Table 2. Compared time of the system

	One-factor authentication	Two-factor authentication
Processing time (s)	4.602	5.5432

5 Conclusions

This research was about the two-factor authentication. The theory in the matter of digital image processing was adapted. To clarify, face position identification as well as the comparison of face codes were applied to OPENCV Program, Java, passwords. The authentication was generated in the form of an application to be tested on Android smartphones. The program could identify the two-factor authentication through which individual's information was scanned and accurately compared with the one stored in the system. The results discovered that the program could recognize face accurately with the mean at 98%, considered as a very high accuracy and at an excellent criteria. Apart from regular access based on usernames and passwords, adding face recognition to the authentication totally consumed 5.5432 s. Despite the mentioned medium speed, the results were satisfactory. Thus, the advantages of the two-factor authentication are that it can protect information from offenders/hackers, and enhance the higher security of users' information.

Acknowledgements. This work was supported by King Mongkut's University of Technology North Bangkok. Contract no.KMUTNB-GEN-59-49.

References

1. Oravec, M.: Feature extraction and classification by machine learning methods for biometric recognition of face and iris. In: Proceedings ELMAR-2014, Zadar, pp. 1–4, (2014)
2. Horiuchi, T., Hada, T.: A complementary study for the evaluation of face recognition technology. In: 47th International Carnahan Conference on Security Technology (ICCST), Medellin, pp. 1–5, (2013)
3. Anggraini, D. R.: Face recognition using principal component analysis and self organizing maps. Student Project Conference (ICT-ISPC), 2014 Third ICT International, Nakhon Pathom, pp. 91–94, (2014)
4. Kafai, M., An, L., Bhanu, B.: Reference face graph for face recognition. IEEE Trans. Inf. Forensics Secur. **9**(12), 2132–2143 (2014)
5. Jian, M., Lam, K.M.: Simultaneous hallucination and recognition of low-resolution faces based on singular value decomposition. IEEE Trans. Circ. Syst. Video Technol. **25**(11), 1761–1772 (Nov 2015)
6. Chou, K.Y., Huang, G.M., Tseng, H.C., Chen, Y.P.: Face recognition based on sparse representation applied to mobile device. In: Automatic Control Conference (CACS), 2014 CACS International, Kaohsiung, pp. 81–86, (2014)
7. Hu J., Peng, L., Zheng, L.: XFace: a face recognition system for android mobile phones. In: IEEE 3rd International Conference on Cyber-Physical Systems, Networks, and Applications (CPSNA), Hong Kong, pp. 13–18, (2015)
8. Voynichka, I. V. and . Megherbi,D. B :Analysis of the effect of using non-composite multi-channel raw color images on face recognition accuracy with arbitrary large off-the-plane rotations. In: IEEE International Symposium on Waltham Technologies for Homeland Security (HST 2015), , MA, pp. 1–6, (2015)
9. Liu, C. :The development trend of evaluating face-recognition technology. In: International Conference on Mechatronics and Control (ICMC 2014), Jinzhou, pp. 1540–1544, (2014)

10. Li, Y. Yang,J., Mengjun, X., Carlson, D., Jang, H. G., Bian, J.: Comparison of PIN- and pattern-based behavioral biometric authentication on mobile devices. In: IEEE Military Communications Conference, (MILCOM 2015), Tampa, FL, pp. 1317–1322, (2015)
11. Meng, W., Wong, D.S., Furnell, S., Zhou, J.: Surveying the development of biometric user authentication on mobile phones. IEEE Commun. Surv. Tutorials **17**(3), 1268–1293 (2015)
12. Lee, J., Oh, Y.: A study on providing the reliable and secure SMS authentication service. In: 2014 IEEE 11th Intl Conf on Ubiquitous Intelligence and Computing and 2014 IEEE 11th Intl Conf on and Autonomic and Trusted Computing, and 2014 IEEE 14th Intl Conf on Scalable Computing and Communications and Its Associated Workshops (UTC-ATC-ScalCom), Bali, pp. 620–624, (2014)
13. Basit, F.-e-., Javed, M.Y., Qayyum, U.: Face recognition using processed histogram and phase-only correlation (POC). In: International Conference on Emerging Technologies (ICET 2007), Islamabad, pp. 238–242, (2007)

Medical Instructional Media in Human Body Systems Using 3D Projection Mapping

Patiyuth Pramkeaw[(⊠)]

Department of Media Technology, King Mongkut's University of Technology Thonburi, Bangkok, Thailand
patiyuth.pra@kmutt.ac.th

Abstract. This study regards the research and develop Medical Instructional Media (MIM) about human's anatomy and 3D Projection Mapping (3DPM) for the application in 3D-MIM by showing the position of human organs and their mechanism, especially digestive and blood circulation systems. The MIM was done by using sensors to detect the movement of audiences to activate media to play video record with the 3DPM. The process of creating MIM consists of data collection, planning and designing the MIM, quality assessment and satisfaction of group sample respectively. From the experimental results, the algorithm works well the height of projector the position. The overall accuracy of uv texture is 89% and the accuracy of 3D mapping is 82%.

Keywords: Medical instructional media · Human body systems · 3D mapping

1 Introduction

Human Anatomy is the subject which studies parts of human body and animals', including organs, organ positions and all related 11 systems. Nowadays, there are several kinds of human anatomy learning media such as lesson book, anatomy book, translated book, CD, VCD, DVD, Clip Video, Poster of the body and the other two-dimensional media [1]. However, these learning media are limited in terms of education due to they cannot respond to the demand of learners when they are insufficient and high-cost. Specially, the subject is very complicated so it is difficult to understand, resulting that students cannot imagine the organ system structure as expected. As a result, they lack motivation to learn while the subject is not attracted as it should be. In consequence, it is really necessary to apply technology to create learning media for human anatomy study.

Regarding the aforementioned problem, the researcher is interested to create learning media by applying 3D Projection Mapping, which has been used in advertising media and public relations for making visual 3D simulation of two human structure models with Blender program. The two systems are blood circulation system and gastrointestinal system [2, 3]. The method is to do the 3D mapping from the model and present on the body model via projector. This leads students to see the 3D picture and organ positions and to understand the two human body system structures clearly and it can be used in the classroom.

© Springer International Publishing AG 2018
T. Theeramunkong et al. (eds.), *Advances in Natural Language Processing,
Intelligent Informatics and Smart Technology*, Advances in Intelligent Systems

2 Preliminary

2.1 3D Projection Mapping

For example, Light Harvest Company has created PR media at Manhattan Bridge by using 3D Projection Mapping under the bridge to produce lights, sounds and motion pictures which were projected from the tunnel which is in the form of a vault so the audiences could watch the picture more realistic [4], And applied 3D Projection Mapping to create lights, sounds and shadows in "Path to the Future" exhibition at University of Sydney in Australia by using the university's building as the screen to present human thinking of materialistic desire. Shades of shadow were used to represent the social classification while the realm was reflected on the building [5] (Fig. 1).

Fig. 1. Shows 3D projection mapping (path to the future exhibition)

LCI Productions Company has created video by using 3D Projection Mapping Demonstrate how to do 3d on mannequins most of this kind of work is done on the structure of the building and using sensors to detect the movement of audiences to activate media to play video record with the 3D Projection Mapping (Fig. 2).

3 Proposed Method

The steps of making learning media about human body system structure with 3D Projection Mapping are as follows: (Fig. 3).

3.1 Measure

Measure size and proportion of the model and record in the program is the first step of creating the learning media so that we can realize the position and basic parts of the visual model for data collection process in the next step.

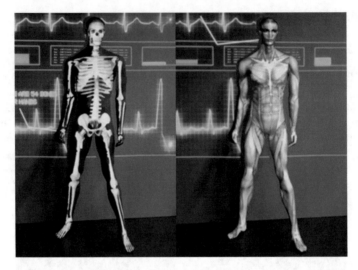

Fig. 2. Shows 3D projection mapping of LCI productions company (3D video projection mapping on mannequin michael)

3.2 Draft Sections the Virtual Model

Do the drafts of every part of the visual model to mark the position in Adobe Illustrator CS6 Program. Due to the researcher needs the picture projected on the model have correct positions [6], such as, curving- bending on the body or the other positions such as facial positions or upper-body positions. Therefore, it requires to be marked. In spot drafting, we should draft important positions and categorize with different colors on the model for the next step (Fig. 4).

3.3 Create Storyboard

Create storyboard to collect content used in the research and to specify necessary resources before creating process. It helps to order the situations sequence roughly.

3.4 Create Texture

Create texture of the visual model in Adobe Photoshop CS6 for more realistic body surface; therefore, we must create simulated skin for the visual model and to add color into the model [9, 10] (Fig. 5).

3.5 Create 3D Model

Create 3D model and set the organ position inside the visual model including animation with Blender 2.73a program. The researcher has created 3D model in 2 systems consisting of gastrointestinal system and blood circulation system. The process started from creating every organ of a system before comparing to the drafting positions and before connecting organs together. Organs positions should be designated in the 3D

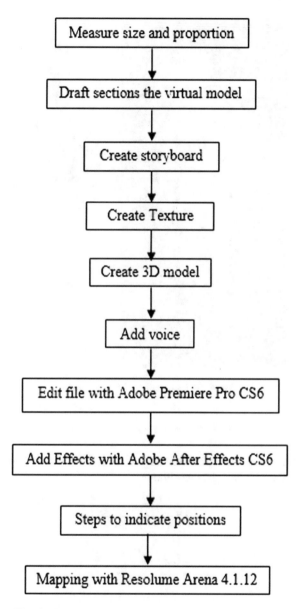

Fig. 3. Flowchart of the steps of making learning media

model [7]. In this step, it requires reference from anatomical theory and animation making for the model in order to increase effects [8]. Because of selection of the two systems, the team must create the two systems more realistic than the 2D pictures. The animation works would be rendered into.PNG files and transformed into.MP4 files with Adobe Premier Pro CS6 to receive the complete video file (Fig. 6).

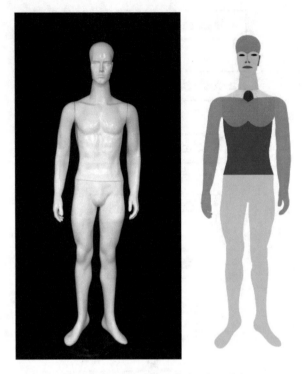

Fig. 4. Spot drafting of visual model

For any point P on the sphere, calculate $\hat{d}$ that being the unit vector form P to the sphere's origin.

Assuming that the sphere's poles are aligned with the Y axis, UV coordinates in the range [0, 1] can then be calculate as follows:

$$u = 0.5 + \frac{\arctan 2(d_z, d_x)}{2\pi} \qquad (1)$$

$$v = 0.5 + \frac{\arcsin(d_y)}{\pi} \qquad (2)$$

3.6 Add Voice

Add voice over of the narrator by human dubbing for the model via recorder application of smartphone, referring the contents of blood circulation system and gastrointestinal system from anatomy book as designated in storyboard and edit the dubbing sound for completeness with Adobe Premier Pro CS6. The file is (.mp3).

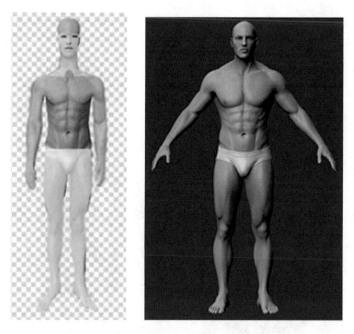

Fig. 5. The original of texture picture which was edited to be the skin for the visual model

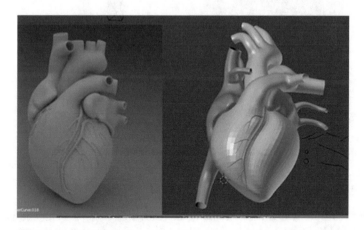

Fig. 6. 3D model (heart) by blender 2.73a and UV texture creation for the model

3.7   Edit File with Adobe Premier Pro CS6

Edit file with Adobe Premier Pro CS6 which is the main program that the researcher had selected for editing video file before adding effects. It starts from adding voice over file (.MP3) as the main character to select picture, 3D animation, video or textures which we had created. The file would be edited referring to the storyboard. Besides, the researchers used Adobe Premier Pro CS6 to add eyes animation (blinking) and mouth (speaking) after the editing process completed which becomes (.MP4) (Fig. 7).

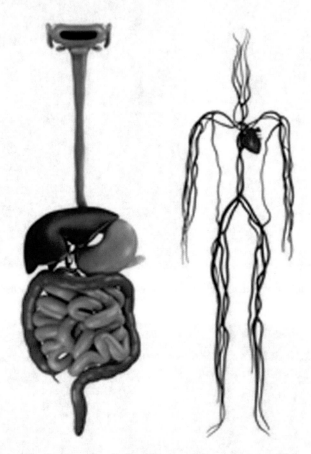

Fig. 7. Example of edited file by adobe premier pro CS6

3.8 Add Effects

Add effects into the work with Adobe After Effects CS6. This technique is to create colors and add lights or effects such as Shine, Particular, Glow and the others into the works, including editing some parts of related videos to increase attraction to capture the audiences. After the effects, the work should be rendered with QuickTime to gain file (.MOV) (Fig. 8).

3.9 Steps to Indicate Positions

Steps to indicate positions between the picture and the visual model including the projection on the visual model via the projector are as follows:

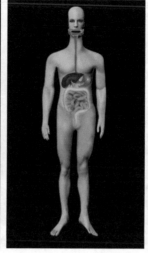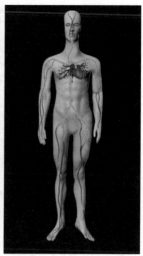

Fig. 8. Example of effects by adobe after effects CS6

(1) Install BenQ MW712 Projector in the computer.
(2) Install visual model and keep suitable length distance with the projector. Use black color background with less light reflection surface or dark because the darkness will enhance the picture quality.
(3) Testing by shooting picture from the projector to the model in order to adjust focus and position.

What we have to be careful to locate position during projection is when all devices have been installed, all positions must be moved until the work is complete because 3D Projection Mapping requires the users to locate fixed positions or scale in order to prevent distortion or damages to the mapping pictures.

3.10   PIR Motion Sensor Step

According to the Fig. 9, the PIR Motion sensor will inspect the motion. The researcher has designed for inspecting movement of people who move around the learning media about body system structure of human. PIR Motion Sensor will inspect the heat from human body and transmit to the Arduino board. There were 2 results which are 1 and 0.1 defines to people passing the area while 0 defines to nobody moves in the area. The data transmission works via Bluetooth HC-05 to send data from PIR Motion Sensor and Arduino to the computer and then connect its circuit with the computer with Microsoft Visual Studio. Connect PIR Motion Sensor with Arduino board by cable and program Code C# to play VDO when inspecting movement, the result from Serial Monitor displays 0 to 1. In this step, the computer will be commanded to play VDO when receiving signal which is the number 1 and will replay in loop when receiving the same result. If the number from Serial Monitor is 0, the computer will do nothing [11].

```
            HC-05 Bluetooth Module by AT

#include <SoftwareSerial.h>
 SoftwareSerial BTSerial(11, 10);
void setup()
{
        pinMode(9, OUTPUT);
        digitalWrite(9, HIGH);
        Serial.begin(9600);
         Serial.println("Enter AT commands:");
        BTSerial.begin(38400);
}

void loop()
{

        if (BTSerial.available())
        Serial.write(BTSerial.read());


        if (Serial.available())
        BTSerial.write(Serial.read());
}
```

Fig. 9. Shows program code with HC-05 bluetooth module by AT

After connecting all circuits, connect with 9 V battery to run the electricity. It will allow PIR Motion Sensor working without using plug. To arrange sensor box and install the model, the team place PIR Motion Sensor connected with Bluetooth HC-05 and battery in the box. The box's cover must be screwed to make holes so that there will be anything against the sensor. After that, place the box in the foot area of the model and turn the sensor to the front.

3.11   Design of Human Anatomy Learning Media with 3D Projection

Before shooting picture for mapping on the model or called 3D Projection Mapping, there must be VDO file which can be done the mapping on the model. The methods are as follows:

– Step1 Texture Creation of the Human Body Model

For virtuality of human anatomy learning media with 3D Projection Mapping, [12, 13] the creation of texture is another importance because the work will become more realistic. The researchers had designed texture of general skin to cover the body, texture of human wearing garment and texture of human body system simulation [14].

– Step2 Add Effects

Effects in the works are various according to the types of pictures in every duration. The effects have been used as texture to add beauty and to highlight specific parts.

– Step3 How to set position among pictures

Length among scene, model and projector was set constantly in order to prevent error during projection [15, 16]. The distance between the body model and the projector is 3.97 m, the height of projector table is 75 cm. and the distance between background and the model is about 20 cm (Fig. 10).

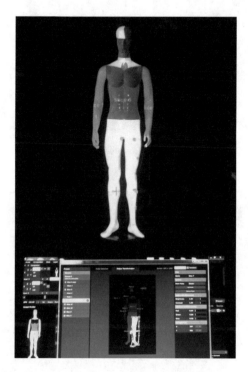

Fig. 10. Example of effects and texture creation by projector

3.12   Mapping

In the final step, record VDO in Camtasia Studio 8 from Resolume Arena which had been already done mapping, the result is VDO file (.MP4) that is able to be projected as designated on the model [17]. The work is called 3D Projection Mapping which will teach anatomical system of human. The content can be divided into 4 major sections which are introduction as in picture (a), the other systems in human body as in picture (b), gastronomical system as in picture (c) and blood circulation system as in picture (d), PIR Motion sensor steps (Fig. 11).

Length among scene, model and projector was set constantly in order to prevent error during projection. The distance between the body model and the projector is 3.97 m., the height of projector table is 75 cm. and the distance between background and the model is about 20 cm (Fig. 12).

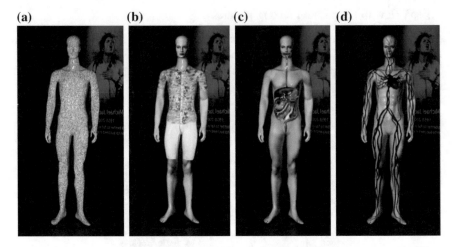

Fig. 11. Picture projected on the model

Fig. 12. Model's and projector's positions

The research purposed to arrange learning media in the form of 3D Projection Mapping about human anatomy by focusing on blood circulation system and gastronomical system. The experimental results are as follows: (Tables 1 and 2).

From the experiments that have tested for 5 level, we found the accuracy of the uv texture and 3D mapping performances are described as follow;

The accuracies from the uv texture capability when the height of projector of 55, 60, 65, 70, 75 are 85, 87, 89, 92, 96, and 89 percents respectively. The accuracies from 3D mapping capability when the height of projector of 55, 60, 65, 70, 75 on are 69, 75, 82, 93, 95, and 82 percents respectively.

Table 1. Experimental results of the performance of the uv texture

	The height of projector the position of the uv texture					Mean
	55	60	65	70	75	
Accuracy	0.85	0.87	0.89	0.92	0.96	0.89

Table 2. Experimental results of the performance of the 3D mapping

	The height of projector the position of the 3D mapping					Mean
	55	60	65	70	75	
Accuracy	0.69	0.75	0.82	0.93	0.95	0.82

4 Conclusion

In this paper, researcher proposed the develop Medical Instructional Media (MIM) about human's anatomy and 3D Projection Mapping (3DPM). From the experimental results, the algorithm works well the height of projector the position. The overall accuracy of uv texture is 89% and the accuracy of 3D mapping is 82%.

References

1. Mao, Y.S.Li.: The design and construction of Human Anatomy Network Course. Anat. Res. **26**, 72–73 (2001)
2. Hsiu-Mei, H., Yi-Chun, T.: Building an interactive and collaboration learning system for virtual reality learning. IEEE. Comput. Symp. **50**, 358–363 (2010)
3. Bhowmik, S.: The parametric museum: combining building information modeling, 3D projection mapping with a community's digital collection for cultural heritage museums. In: IEEE Digital Heritage International Congress, vol. 2, pp. 449 (2013)
4. Drive Productions, Ltd, Mapping demos will blow your mind, 2014 [online], Available: http://www.creativebloq.com/video/projection-mapping-912849
5. Creativebloq, Mapping demos will blow your mind, 2014 [online], Available: http://www.creativebloq.com/video/projection-mapping-912849
6. Chen, H., Dong, M.: 3D map building based on projection of virtual height line. In: IEEE Circuits and Systems, pp. 1822–1825 (2008)
7. Saez, J.M., Escolano, F.: A global 3D map-building approach using stereo vision, In: Proceedings of the 2004 IEEE International Conference on Robotics and Automation, pp. 1197–1202 (2004)
8. Matteo, D., Ricardo, M., Marco, C., Paolo, C., Roberto, S.: Flow-based local optimization for image-to-geometry projection. IEEE Trans. Visual Comput. Graphics **18**(3), 463–474 (2012)
9. Xuliang, G., Yuhui, H., Baoquan, S.: Texture mapping based on projection and viewpoint. In: IEEE Digital Home (ICDH), pp. 173–179 (2014)

10. Zhang, E., Mischaikow, K., Turk, G.: Feature-based surface parameterization and texture mapping. ACM TOG **24**(1), 1–27 (2005)
11. Smith, R.T., Thomas, B.H., Piekarski, W.: Text note: digital foam. In Proceedings of IEEE Symposium on 3D User Interfaces (3DUI), pp. 35–38 (2008)
12. Xin, S., Guofu, X., Yue, D., Baining, G.: Diffusion curve textures for resolution independent texture mapping. ACM SIGGRAPH conference proceedings (2012)
13. Xuliang, G., Yuhui H., Baoquan Z., Shujin L.: Texture mapping based on projection and viewpoints, international conference on digital home. In: International Conference on Digital Home of the IEEE Computer Society, pp. 173–179 (2014)
14. Yao, J., Zeyun, S., Jun, S., Jin, H., Ruofeng, T.: Content-aware texture mapping. Computational Visual Media conference, pp. 152–161 (2014)
15. De Boer, C.N., Verleur, R., Heuvelman, A., Heynderickx, I.: Added value of an autostereoscopic multiview 3-d display for advertising in a public environment. Displays. **31**(1), 1–8 (2010)
16. Nisula, P., Kangasoja, J., Karukka, M.: Creating 3D projection on tangible objects In Proceedings of IEEE Virtual Reality, pp. 187–191 (2013)
17. Alazawi, E., Aggoun, A., Abdul Fatah, O., Abbod, M., Swash, M.R.: adaptive depth map estimation from 3D integral image. In: IEEE International Symposium Broadband Multimedia Systems Broadcasting, pp. 1–6, London, UK, (2013)

Development of Electromagnetic Wave Propagation in Microstrip Antenna Using the Novel Finite Element Method

Nattapon Jaisumroum[✉] and Somsak Akatimagool

Department of Teacher Training in Electrical Engineering, Faculty of Technical Education, King Mongkut's University of Technology North Bangkok, 1518 Pracharat 1 Road, Wongsawang, Bangsue, Bangkok 10800, Thailand
nattaponj89@gmail.com, ssa@kmutnb.ac.th

Abstract. This paper presents a novel method of numerical techniques using software development by applying the finite element method for solving an electromagnetic engineering problem. We introduce to analyze for electromagnetic field and wave propagation. This is to reduce the education and the research budget due to the reason that, in the present, the packaging software is very expensive and cannot adjust the source code so that not enough to simulation by research. This study synthesis and design of complex electromagnetic circuits for solving more complicated electromagnetic problem. The simulation results showed in this report that our program developed from MATLAB gives the proper results that can be used for education and solving any more complicated electromagnetic problem.

Keywords: Finite element method · Electromagnetic simulation · Microstrip electromagnetic wave propagation

1 Introduction

Presently, the communication system is important in the technology development. Usually the kind of communication system requires high frequency or microwave, which are often used and developed in communication system technology. In education and research, the design and synthesis of the microwave are accomplished by numerical methods [1, 2]. Several computational methods related to electromagnetic and wave propagation have been extracted, such as the Method of Moment (MoM) [3, 4], the Finite Difference Time Domain (FDTD) [5–11], and the finite element method (FEM) [12–17] which is one of the best known strategies for the arrangement of halfway differential comparisons on joined computational mechanics and applied mathematics. It is a technique for clearing up a differential comparison subject to certain limit values. Furthermore previously, its cutting edge structure started in the field of structural mechanics throughout the late 1950s; the primary particular utilization of the expression "element" will be because of no lesser a man than Courant. In basic with the MoM, its recorded antecedents would a long way more seasoned over this, for this the event going back to the nineteenth century; the vibrational techniques

T. Theeramunkong et al. (eds.), *Advances in Natural Language Processing,*
Intelligent Informatics and Smart Technology, Advances in Intelligent Systems

were initially portrayed by Lord Rayleigh. The techniques have been broadly and routinely utilized as a part of structure mechanics, and additionally to computational fluid dynamics, computational thermodynamics, the numerical arrangement of Schrö-dinger's equation, field issues in general, and of course, in electromagnetic. With the foundation, with the FDTD also MoM, onlooker will distinguish numerous charac-teristics in a similar manner as both of these techniques are in the treatment to take after; in reality, they will likely not be astonished to realize that every one of the three can be planned inside of a weighted remaining setting (although how to do this for the FDTD is not obvious). In a similar manner as the MoM, the center thought is to displace some obscure capacity on a domain by gathering elements, referring to shape, however obscure plentiful. Different with the individuals crucial FDTD, the place of the close estimation of the E, also H fields, will be ceaselessly done with admiration to a rectangular, staggered grid, the FEM permits greatly all geometrical segments once a chance to be used and (usually) main one-grid utilization. The most generally utilized elements are known as simplified this just means line elements in 1D, triangular in 2D, and tetrahedral in 3D. Nonetheless, rectangular, kaleidoscopic and even curvilinear elements, likewise, are found across the board application. Subsequent to the enhanced geometrical modeling made conceivable particularly by triangular or tetrahedral meshes, one of the real characteristics recognizing the FEM from the FDTD, our investigation of the FEM will be to provide a great extent confined to these element. Intrigued peruses might discover treatments of other elements' shapes in the references. Like the FDTD, however not at all like the MOM, the FEM depends on a local description of the field quantities, determined from the differential equations, the Maxwell's equations, and does not consequently join the Somerfield radiation condi-tion. On practice, this intends a few structure of mesh termination scheme is required.

In this paper using the Finite Element Method (FEM) for solve the problem of simulation for electromagnetic and wave propagation in microstrip antenna. The analysis model based on Finite Element Method Theory was implemented by using the Graphic User Interface (GUI) function of MATLAB software. User can define the dimensions of the conducting plate and selected the shape of patch conductor also cavity dimension and number of mode for analysis of the electromagnetic wave in the spectral domain of conducting plate. Therefore, the developments of numerical methods are important for an efficient electromagnetic simulation tool. In this study, we introduce an official electromagnetic simulation tool for analysis of electromagnetic wave propagation in conducting cavity using Finite Element Method.

2   Theory

2.1   Wave Propagation

The computation for the electromagnetic wave propagation consists of ampli-tude, direction of the incident, reflected and transmitted waves that propagation in a shield. The waves can be calculated in the real domain and spectrum domain scattering in waveguide concern about frequency of waveguide concern about frequency of the wave in TE and TM mode by spectral domain, the conductivity shown in Fig. 1.

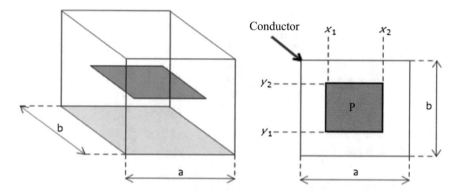

Fig. 1. Conductor patch in conductor box

For Finite Element method, the conductor patch can be analyzed in many shapes on the conductor patch; this is the advantage of this method. The shape of conductor patch in this study can be replaced by any shapes in Fig. 2.

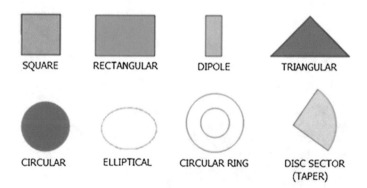

SQUARE RECTANGULAR DIPOLE TRIANGULAR

CIRCULAR ELLIPTICAL CIRCULAR RING DISC SECTOR (TAPER)

Fig. 2. Shape of conductor patch

Electric fields for Transverse Electric (TE) mode by following equation is

$$E_x^{TE} = \frac{n}{b\sqrt{\frac{n^2}{b^2} + \frac{m^2}{a^2}}} \sqrt{\frac{2\tau_{mn}}{ab}} \cos\left(\frac{m\pi x}{a}\right) \sin\left(\frac{n\pi y}{b}\right)$$

$$E_y^{TE} = \frac{-m}{a\sqrt{\frac{n^2}{b^2} + \frac{m^2}{a^2}}} \sqrt{\frac{2\tau_{mn}}{ab}} \sin\left(\frac{m\pi x}{a}\right) \cos\left(\frac{n\pi y}{b}\right)$$

(1)

Electric fields for Transverse Magnetic (TM) mode by following equation is

$$E_x^{TE} = \frac{n}{b\sqrt{\frac{n^2}{b^2} + \frac{m^2}{a^2}}} \sqrt{\frac{2\tau_{mn}}{ab}} \cos\left(\frac{m\pi x}{a}\right) \sin\left(\frac{n\pi y}{b}\right)$$

$$E_y^{TE} = \frac{-m}{a\sqrt{\frac{n^2}{b^2} + \frac{m^2}{a^2}}} \sqrt{\frac{2\tau_{mn}}{ab}} \sin\left(\frac{m\pi x}{a}\right) \cos\left(\frac{n\pi y}{b}\right)$$

(2)

While, $\tau_{mn} = 2$ if m and n not equal zero
$\tau_{mn} = 1$ if m and n equal zero

2.2   The Boundary Value Problem of Microstrip

The open stub microstrip type can compute the analysis of boundary problem's exact solution for half contact to the air kind and half contact to the conductor patch. In the operation it can be analysed by waveguide equation shown in Fig. 3. It also closes loop like waveguide. The optimal length "a" and "b" must be more than 10 folds of patch length.

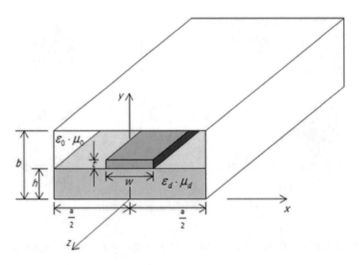

Fig. 3. Microstrip line

2.3   Finite Element Method

The finite element method can solve the problems and provides solutions in many dimensions, such as the elements shown in Fig. 4.

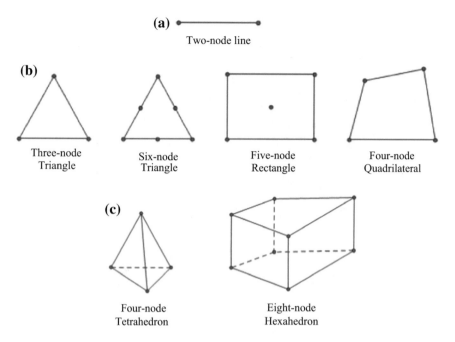

Fig. 4. Elements **a** one-dimension, **b** two-dimension, **c** three-dimension

There are 5 steps in Finite Element Method

1. Pre-computation step

The preprocessing step requires the automatic mesh generator. It subdivides the region under learn into a set of elements, and usually most triangles to fit a generic two-dimensional shape. The mesh generator creates following information about the mesh shown in Fig. 5.

The mesh data must be computed by the LAPLACE to solve the problem properly, as contained in Tables 1, 2 and 3. Table 1 shows the list of the node coordinates and Table 2 shows the list of the nodes are Diriclet boundary conditions, while Table 3 shows the entire element and describes the nodes.

2. Building Element Matrices

The unknown function $u(x, y)$ approximates each element using the polynomial expression.

$$u(x, y) = a + bx + cy = \begin{bmatrix} 1 & x & y \end{bmatrix} \begin{bmatrix} a \\ b \\ c \end{bmatrix} \tag{3}$$

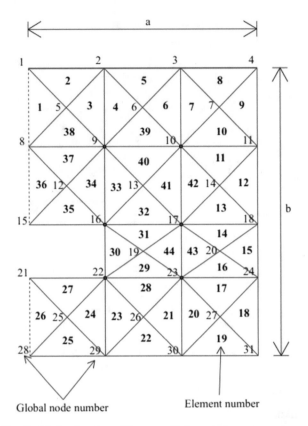

Fig. 5. Mesh of region of interest. Node and element are shown

Table 1. Node coordinates

Node number	x	y	Node number	x	y	Node number	x	y
1	0.000	1.000	12	0.167	0.580	23	0.667	0.280
2	0.333	1.000	13	0.500	0.580	24	1.000	0.280
3	0.667	1.000	14	0.833	0.580	25	0.167	0.140
4	1.000	1.000	15	0.000	0.440	26	0.500	0.140
5	0.167	0.860	16	0.333	0.440	27	0.833	0.140
6	0.500	0.860	17	0.667	0.440	28	0.000	0.000
7	0.833	0.860	18	1.000	0.440	29	0.333	0.000
8	0.000	0.720	19	0.500	0.360	30	0.667	0.000
9	0.333	0.720	20	0.833	0.360	31	1.000	0.000
10	0.667	0.720	21	0.000	0.280			
11	1.000	0.720	22	0.333	0.280			

Table 2. Dirichlet boundary conditions

Node number	Label	Prescribed potential	Node number	Label	Prescribed potential
1	1	0.000	21	2	1.000
2	1	0.000	22	2	1.000
3	1	0.000	24	1	0.000
4	1	0.000	28	1	0.000
11	1	0.000	29	1	0.000
15	2	1.000	30	1	0.000
16	2	1.000	31	1	0.000
18	1	0.000			

Table 3. Connection between global and local numbering schemes

El.		Local node number			El.		Local node number			El.		Local node number		
No.	Lb.	1	2	3	No.	Lb.	1	2	3	No.	Lb.	1	2	3
1	1	5	1	8	16	1	20	23	24	31	1	16	19	17
2	1	5	2	1	17	2	23	27	24	32	1	13	16	17
3	1	5	9	2	18	2	24	27	31	33	1	9	16	13
4	1	6	2	9	19	2	27	30	31	34	1	9	12	16
5	1	3	2	6	20	2	23	30	27	35	1	12	15	16
6	1	6	10	3	21	2	23	26	30	36	1	8	15	12
7	1	7	3	10	22	2	26	29	30	37	1	8	12	9
8	1	7	4	3	23	2	22	29	26	38	1	5	8	9
9	1	7	11	4	24	2	22	25	29	39	1	6	9	10
10	1	7	10	11	25	2	25	28	29	40	1	9	13	10
11	1	14	11	10	26	2	21	28	25	41	1	10	13	17
12	1	14	18	11	27	2	21	25	22	42	1	10	17	14
13	1	14	17	18	28	1	22	26	23	43	1	17	23	20
14	1	17	20	18	29	1	19	22	23	44	1	17	19	23
15	1	18	20	24	30	1	16	22	19					

The Cartesian coordinate approximated used function potential assumes the value of triangle node.

$$\begin{cases} u_1(x_1, y_1) = a + bx_1 + cy_1 \\ u_2(x_2, y_2) = a + bx_2 + cy_2 \\ u_3(x_3, y_3) = a + bx_3 + cy_3 \end{cases} \tag{4}$$

Recast the equation to easy express.

$$\begin{bmatrix} a \\ b \\ c \end{bmatrix} = \begin{bmatrix} 1 & x_1 & y_1 \\ 1 & x_2 & y_2 \\ 1 & x_3 & y_3 \end{bmatrix}^{-1} \begin{bmatrix} u_1 \\ u_2 \\ u_3 \end{bmatrix} \qquad (5)$$

Introducing (5) into (3) yields and A is the area of element.

$$u(x, y) = \frac{1}{2A} [1 \quad x \quad y] \cdot \begin{bmatrix} x_2y_3 - x_3y_2 & x_3y_1 - x_1y_3 & x_1y_2 - x_2y_1 \\ y_2 - y_3 & y_3 - y_1 & y_1 - y_2 \\ x_3 - x_2 & x_1 - x_3 & x_2 - x_1 \end{bmatrix} \cdot \begin{bmatrix} u_1 \\ u_2 \\ u_3 \end{bmatrix} \qquad (6)$$

It can be shown that linear shape functions on a triangle with the simplex coordinates.

$$\alpha_1 = \frac{1}{2A}[(x_2y_3 - x_3y_2) + (y_2 - y_3)x + (x_3 - x_2)y]$$

$$\alpha_2 = \frac{1}{2A}[(x_3y_1 - x_1y_3) + (y_3 - y_1)x + (x_1 - x_3)y] \qquad (7)$$

$$\alpha_3 = \frac{1}{2A}[(x_1y_2 - x_2y_1) + (y_1 - y_2)x + (x_2 - x_1)y]$$

Show in Fig. 6 and which have the two following properties

$$\alpha_i = \begin{cases} 1, i = j \\ 0, i \neq j \end{cases} \qquad (8)$$

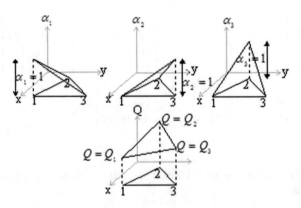

Fig. 6. Shape functions and linear approximation of element

The functional elements can be expressed as shown in (9).

$$W^{(e)} = \frac{1}{2} \in^{(e)} \sum_{i=1}^{3} \sum_{j=1}^{3} u_1 u_j \int_{\Delta} (e) \nabla_t \alpha_i \cdot \nabla_t \alpha_j ds \tag{9}$$

Recasting the equation in matrix form.

$$W^{(e)} = \frac{1}{2} \in^{(e)} [U]^t \left[S^{(e)} \right] [U] \tag{10}$$

With

$$[U] = \begin{bmatrix} u_1 \\ u_2 \\ u_3 \end{bmatrix} \quad [S^{(e)}] = \begin{bmatrix} s_{11}^{(e)} & s_{12}^{(e)} & s_{13}^{(e)} \\ s_{21}^{(e)} & s_{22}^{(e)} & s_{23}^{(e)} \\ s_{31}^{(e)} & s_{32}^{(e)} & s_{33}^{(e)} \end{bmatrix} \tag{11}$$

While; $S_{21}^{(e)} = S_{12}^{(e)}, \quad S_{31}^{(e)} = S_{13}^{(e)}, \quad S_{32}^{(e)} = S_{23}^{(e)}$

$S_{11}^{(e)} = \frac{1}{4A} \left[(y_2 - y_3)^2 + (x_3 - x_2)^2 \right],$ $\qquad S_{12}^{(e)} = \frac{1}{4A} [(y_2 - y_3)(y_3 - y_1) + (x_3 - x_2)(x_1 - x_3)],$

$S_{13}^{(e)} = \frac{1}{4A} [(y_2 - y_3)(y_1 - y_2) + (x_3 - x_2)(x_2 - x_1)],$ $S_{22}^{(e)} = \frac{1}{4A} \left[(y_3 - y_1)^2 + (x_1 - x_3)^2 \right],$

$S_{23}^{(e)} = \frac{1}{4A} [(y_3 - y_1)(y_1 - y_2) + (x_1 - x_3)(x_2 - x_1)],$ $S_{33}^{(e)} = \frac{1}{4A} \left[(y_1 - y_2)^2 + (x_2 - x_1)^2 \right],$

3. Assembling the Global Matrix

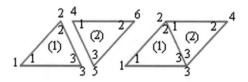

Fig. 7. Two-triangle meshes with local and global node numbering

As shown in Fig. 7, it contains the unknowns of the problem which the equation can be written as (12)

$$W = W^{(1)} = W^{(2)} = \frac{1}{2} \in^{(1)} \left[U_g^{(1)} \right]^t \left[S^{(1)} \right] \left[U_g^{(1)} \right] + \frac{1}{2} \in^{(2)} \left[U_g^{(2)} \right]^t \left[S^{(2)} \right] \left[U_g^{(2)} \right]$$

$$= \frac{1}{2} [U]^t [S][U] \tag{12}$$

where $[S]$ is the global 4×4 FEM Matrix.

$$[S] = \in^{(1)} \begin{bmatrix} S_{11}^{(1)} & S_{12}^{(1)} & S_{13}^{(1)} & 0 \\ S_{21}^{(1)} & S_{22}^{(1)} & S_{23}^{(1)} & 0 \\ S_{31}^{(1)} & S_{32}^{(1)} & S_{33}^{(1)} & 0 \\ 0 & 0 & 0 & 0 \end{bmatrix} + \in^{(2)} \begin{bmatrix} 0 & 0 & 0 & 0 \\ 0 & S_{11}^{(2)} & S_{13}^{(2)} & S_{12}^{(2)} \\ 0 & S_{31}^{(2)} & S_{33}^{(2)} & S_{32}^{(2)} \\ 0 & S_{21}^{(2)} & S_{23}^{(2)} & S_{22}^{(2)} \end{bmatrix} \tag{13}$$

4. Minimizing the Functional to avoid unnecessary complex nodes to the matrix of linear equation

$$\begin{bmatrix} 1 & 0 & 0 & 0 \\ 0 & S_{22} & S_{23} & 0 \\ 0 & S_{32} & S_{33} & 0 \\ 0 & 0 & 0 & 1 \end{bmatrix} \begin{bmatrix} u_1 \\ u_2 \\ u_3 \\ u_4 \end{bmatrix} = \begin{bmatrix} \overline{u}_1 \\ -S_{11}\overline{u}_1 - S_{14}\overline{u}_4 \\ -S_{41}\overline{u}_1 - S_{44}\overline{u}_4 \\ \overline{u}_4 \end{bmatrix} \tag{14}$$

5. Post processing

As mentioned, the minimum value assumed for $u(x, y) = \overline{u}(x, y)$ represents the stored energy $\overline{W}$ in the FEM computation shown in (15)

$$\overline{W} = \frac{1}{2} [U]^t [S] [U] \tag{15}$$

3   Electromagnetic Wave Propagations Simulation Design

The electromagnetic wave propagation simulation created through a graphical user interface (GUI) function and computation of MATLAB program and electromagnetic wave propagation simulation shown in Fig. 8.

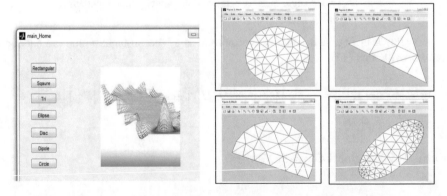

Fig. 8. Electromagnetic wave propagation simulation program

4   Results

The developed electromagnetic wave propagation simulation program can compute and show an accurate graph pattern of electric field and magnetic field same the theory result, including the graph pattern of TE mode and TM mode, shown in Figs. 9 and 10. The advantage of this research can analyze many patch shape of wave propagation. So the present of approached finite element method can be applied in design and can analyze microstrip antenna, also electromagnetic engineering.

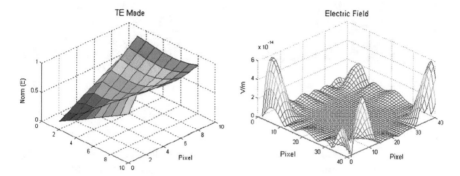

Fig. 9. Electric wave and electric field in TE mode

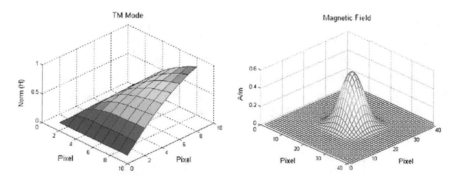

Fig. 10. Magnetic wave and magnetic field in TM mode

The Figure 11 shows the electric and magnetic field for the use of configuration in any mode, such as M-Mode and N-Mode which are 3 × 3, so the modes are 9 numbers of the modes. It shows as density's electric field of the boundary circle conductivity plate area, it is adjusted by the wavelength number. Furthermore, the density's magnetic field depends on the maximal conductivity on the plate area (located in the middle). In the configuration, M and N modes are 15 × 15, and there are 225 modes experimented.

Electric Field

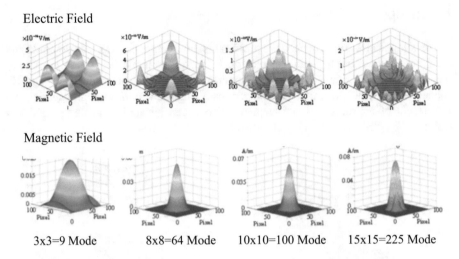

Magnetic Field

3x3=9 Mode   8x8=64 Mode   10x10=100 Mode   15x15=225 Mode

Fig. 11. Electric (TE) and Magnetic (TM) Field in each mode

We consider a rectangular waveguide. The waveguide is assumed to be homogeneously filled with a material of permittivity $\varepsilon = \varepsilon_r \varepsilon_0$ and permeability $\mu = \mu_r \mu_0$. For TE polarization, F is an electric field E_t, whereas for TM polarization F_t is the magnetic field H_t. We are interested in determining the eigenvalues γ^2 from the transverse propagation constants $\beta = \sqrt{\gamma^2 - \varepsilon_r \mu_r k_0^2}$. Assembly of the element equations yields the global matrix equation system.

The FEM matrix is used in computing the known capacitance of a given transmission line, such as the shielded microstrip line in Fig. 3. In this case, the wavenumber k_0 is set to zero and thus only the $[K_\nabla^e]$ matrix is used to evaluate the capacitance C_0. A situation, where both sub-matrices $[K_\nabla]$ and $[K]$ obtained from the assembly of $[K_\nabla^e]$ and $[K^e]$ given in Eq. 16, are used for computing the waveguide cutoff wavenumbers. The cutoff wavenumbers are obtained by solving the generalized eigenvalue problem

$$-[K_\nabla]\{H_z\} = \gamma^2 [K]\{H_z\}$$

(16)

Table 4. Analytical and numerical γa values for TE modes

Mode	Analytical result	FEM result
TE_{10}	3.14	3.14
TE_{20}	6.28	6.28
TE_{01}	6.28	6.28
TE_{11}	7.02	7.02
TE_{21}	8.88	8.88
TE_{30}	9.42	9.42

Table 5. Analytical and numerical γa values for TM modes

Mode	Analytical result	FEM result
TM$_{11}$	7.02	7.02
TM$_{21}$	8.88	8.88
TM$_{31}$	11.33	11.33
TM$_{12}$	12.95	12.95
TM$_{22}$	14.05	14.04
TM$_{32}$	15.71	15.70

In this case, γ^2 is an eigenvalue and γ refers to the cutoff wavenumbers. These calculations were carried out eigenvalue by

$$\gamma = \sqrt{\left(\frac{m\pi}{a}\right)^2 + \left(\frac{n\pi}{a/2}\right)^2} \tag{17}$$

In FEM. for TE and TM polarization in the rectangular waveguide with analytical and numerical values for γa are compared in Tables 4 and 5 [18]. The analytical and numerical values as seem.

5 Conclusion

This study synthesis and design of complex electromagnetic circuits can be implemented for solving more complicated electromagnetic problems. The simulation results in this report shown that our program developed from MATLAB offers the proper results and can be used in engineering education and solving high-complexity electromagnetic problems.

References

1. Zhao, B., Young, J.C., Gedney, S.D.: SPICE lumped circuit subcell model for the discontinuous Galerkin finite-element time-domain method. Microwave Theory Tech. IEEE Trans. **60**(9), 2684–2692 (2012)
2. Peng, Z., Lim, K.H., Lee, J.F.: A discontinuous Galerkin surface integral equation method for electromagnetic wave scattering from nonpenetrable targets. Antennas Propag. IEEE Trans. **61**(7), 3617–3628 (2013)
3. Liu, Z.L., Wang, C.F.: Efficient iterative method of moments—physical optics hybrid technique for electrically large objects. Antennas Propag. IEEE Trans. **60**(7), 3520–3525 (2012)
4. Gruber, M.E., Eibert, T.F.: Simulation of reverberation chambers using method of moments with cavity Green's function and spectral domain factorization. In: 2013 IEEE International Symposium on Electromagnetic Compatibility (EMC), pp. 808–812, August 2013
5. Aodsup, K., Kulworawanichpong, T.: Simulation of lightning surge propagation in transmission lines using the FDTD Method. World Acad. Sci. Eng. Technol. **71** (2012)

6. Cakir, G., Sevgi, L., Ufimtsev, P.Y.: FDTD modeling of electromagnetic wave scattering from a wedge with perfectly reflecting boundaries: comparisons against analytical models and calibration. Antennas Propag. IEEE Trans. **60**(7), 3336–3342 (2012)
7. Francés, J., Pérez-Molina, M., Bleda, S., Fernández, E., Neipp, C., Beléndez, A.: Educational software for interference and optical diffraction analysis in Fresnel and Fraunhofer regions based on MATLAB GUIs and the FDTD method. Educ. IEEE Trans. **55**(1), 118–125 (2012)
8. Zaytsev, K.I., Gorelik, V.S., Khorokhorov, A.M., Yurchenko, S.O.: FDTD simulation of the electromagnetic field surface states in 2D photonic crystals. In Journal of Physics: Conference Series, vol. 486, no. 1, p. 012003. IOP Publishing (2014)
9. Toroglu, G., Sevgi, L.: Finite-difference time-domain (fdtd) matlab codes for first-and second-order em differential equations [testing ourselves]. Antennas Propag. Mag. IEEE **56**(2), 221–239 (2014)
10. Gaffar, M., Jiao, D.: An explicit and unconditionally stable FDTD method for electromagnetic analysis. Microwave Theory Tech. IEEE Trans. **62**(11), 2538–2550 (2014)
11. Nguyen, B.T., Furse, C., Simpson, J.J.: A 3-D stochastic FDTD model of electromagnetic wave propagation in magnetized ionosphere plasma. Antennas Propag. IEEE Trans. **63**(1), 304–313 (2015)
12. Ham, S., Bathe, K.J.: A finite element method enriched for wave propagation problems. Comput. Struct. **94**, 1–12 (2012)
13. Li, J., Huang, Y.: Time-domain finite element methods for Maxwell's equations in metamaterials, vol. 43. Springer Science & Business Media (2012)
14. Zhong, L., Chen, L., Shu, S., Wittum, G., Xu, J.: Convergence and optimality of adaptive edge finite element methods for time-harmonic Maxwell equations. Math. Comput. **81**(278), 623–642 (2012)
15. Li, J., Huang, Y., Yang, W.: An adaptive edge finite element method for electromagnetic cloaking simulation. J. Comput. Phys. **249**, 216–232 (2013)
16. Jiang, X., Li, P., Zheng, W.: Numerical solution of acoustic scattering by an adaptive DtN finite element method. Commun. Comput. Phys. **13**(05), 1277 (2013). 1244
17. Anjam, I., Valdman, J.: Fast MATLAB assembly of FEM matrices in 2D and 3D: Edge elements. Appl. Math. Comput. **267**, 252–263 (2015)
18. Volakis, J.L., Chatterjee, A., Kempel, L.C.: Finite Element Method for Electromagnetics: antennas, microwave circuits, and scattering applications, vol. **6**, pp. 144–145. John Wiley & Sons (1998)

Detection of Diabetic Retinopathy Using Image Processing

Ratikanlaya Tanthuwapathom and Narit Hnoohom[(⊠)]

Image, Information and Intelligence Laboratory, Department of Computer Engineering, Faculty of Engineering, Mahidol University, Nakhon Pathom, Thailand
t.ratikanlaya@gmail.com, narit.hno@mahidol.ac.th

Abstract. Diabetic retinopathy has been found in from 22% of diabetic patients from the latest survey, which can lead to blindness later. Thus, eye exams should be scheduled no less than annually for patients with diabetes. On the other hand, there is an apparent deficiency in the number of medical professionals specializing in ophthalmology, which may hinder the discovery and care of diabetic retinopathy. The idea to stimulate a detection system for screening of diabetic retinopathy by using morphological and image segmentation methods to facilitate and make a preliminary decision with ophthalmologists is introduced. From the experimental results, the detection system provides good exudate detection and classification results for all 60 retina images for accurate up to 85%.

Keywords: Exudate · Optic disc · Retina image · Microaneurysms · Retinal hemorrhages

1 Introduction

At present, diabetic retinopathy has been found 3.5 million patients in Thailand. It also found that a person with diabetes losing eyes possibility is up to 20% more than ordinary people. The studies found that patients with less than 10 years of diabetes have the risk of having the diabetic retinopathy at 7%. However, the risk would increase to 63% for patients with more than 15 years of diabetes. Anyway, patients with well glucose control can still have diabetic retinopathy in the longer term or older age. Therefore, patients with diabetes should be checked their optical health at least once a year. Therefore, the detection of abnormalities in the retina of diabetic patients is significant for treatment. However, in Thailand there are not enough ophthalmologists for diabetic patients. Therefore, creating disease diagnosing system will allow doctors to work more quickly and help patients not to lose their eyes according to the delayed detection of diabetic retinopathy conditions. Types of lesion we found were such as microaneurysms, retinal hemorrhages and exudates [1]. Several exudate detection methods have been proposed.

Ahmad et al. [1] exhibited the expansion of a segmented discovery and categorization scheme, in which the DR procedure is contingent on the number of exuded substances. Exudation identification involves two primary procedures: rough segmentation and fine segmentation. Utilization of a morphological function and column-wise neighborhoods process comprises rough segmentation, while a good classification should be done by the morphological reconstruction. The presented method was the using of retinal image database from Sungai Buloh Hospital in Malaysia, together with more appropriate retina feature extractions. This classification method would also be a simple and fast basis. Result of the system test was 60% of correct of classification.

The prompt discovery of diabetic retinopathy employing green and red channels was recommended by Sreng et al. [2]. Image binarization, region of interest (ROI) based separation and morphological reconstructions are included in the exudate extraction. Result of the system test was 91% of correct of classification.

A process for the exposure of exudates in retina imaging utilizing morphological and candy methods was offered by Eman Shahin et al. [3]. Microaneurysm discovery employing histogram equalization, morphology and a candy-edge indicator is subsequent. Result of the system test was 92% of correct of classification by artificial neural network (ANN).

Categorization of average and non-proliferative phases of diabetic retinopathy (NPDR) through identifying the quantity of blood vessels and the amount of blood in retinal images was recommended by Verma et al. [4]. The variance in the vascular and background could be catalogued by blood vessels in the retinal image. Thresholding comprised the main separation method for this study, while the exudate detector employed a dot-blot hemorrhage microaneurysm (MA). Result of the system test was 90% of correct of classification.

A process for utilizing a computer to monitor the retinopathy stage with color retina images was reported by Du Ning and Li Yafen [5]. In retinal images used for feature assessment, the methods of morphological processing and texture analysis were used. The areas of vascular, exudates, acuity of identical areas, and features to be supplemented to the support vector machine (SVM) comprise various examples. The result showed that the correct classification of diabetic retinopathy at 93%.

This paper presents the detection of diabetic retinopathy based on morphological and image segmentation methods that is able to detect the thick greasy substance in the retinal image. Furthermore, this method does not even require a circular in optic disc detection process that is able to apply the proposed method on others database of diabetic retinopathy. In the first of proposed method is detect and subtract the optic disk with morphological method to avoid noise in the exudate detection. After eliminating optic disk from the image of retina, the thick greasy substance was extracted. To achieve high efficiency, we executed exudate detection due to the different color to the retinal image. Only green channel of RGB color model was used for detection. Then, the green channel would be segmented by adaptive thresholding process. A good segmentation based on an appropriate thresholding selection.

The rest of paper is organized as follows: the introduction and related works are presented in the Sect. 1. In Sect. 2 describes the proposed method used for detecting, extracting, and classifying the thick greasy substance in the retinal images. The experimental results and discussions are described in Sect. 3. Finally, the conclusion are presented in the Sect. 4.

2   Methodology

In this section, the methodology for exudate detection is consisted of four main stages: image acquisition, pre-processing, exudate extraction, and exudate classification (Fig. 1).

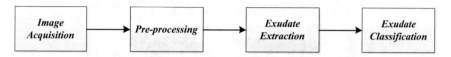

Fig. 1. The procedure of the proposed exudates classification

2.1   Image Acquisition

In the first phase of the exudate detection is the image acquisition. The retina images were collected from Institute of Medical Research and Technology Assessment. The image data were converted in the JPEG file format with size 813 × 499, 60 images in totals, 15 images as a normal retina images and 45 images as an abnormal retina images.

2.2   Pre-processing

This pre-processing is to prepare the image data for using in a detection of diabetic retinopathy using image processing.

2.2.1   RGB Color Model
RGB color model is used to analyze lesions. Only green channel was used because this channel could keep most elements of an image with high contrast between blood vessels and optic disk when compared to red channel and blue channel [2–4, 6] as shown in Fig. 2.

2.3   Exudate Extraction

Exudate Extraction is composed of two stages: the process of separating background from retinal images and optical disc detection.

2.3.1   The Process of Separating Background from Retinal Images
This process is to separate the image background with the local thresholding segmentation by separating the background from the object of interest into small parts. All parts would be applied by the same thresholding value to separate objects from the background as shown in Fig. 3.

2.3.2   Optical Disc Detection
This process is to erase the retina from the image, leaving only the exudates by morphological [5, 7, 8]. Closing or expanding the image structure in order to make the

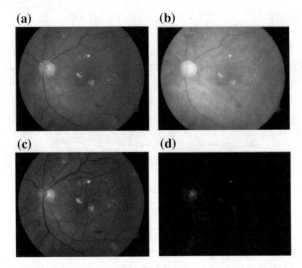

Fig. 2. a Original image **b** Red channel **c** Green channel **d** Blue channel

Fig. 3. Segmentation image using local thresholding with green channel

image larger. The looks of expanded images depended on the structuring element used to scan on the image. After the dilation, the next process was the erosion to reduce the object size. The looks of eroded images depended on the structuring element used to scan on the image by bringing the expanded image through dilation process to the Erosion process to eliminate the exudates, leaving only the retina. After the closing process, the images would be brought through dilation again to increase the retina size. Then, the images that passed the local thresholding process would be used to delete the image of the retina. As a result, the retina on the image would disappear, leaving only the exudates as shown in Fig. 4.

In this paper, we used 4-connectivity based on connected components analysis to define the adjacent pixels for finding the exudates, processing by viewing areas with four connecting points. Objects failing to meet criteria will be cut out of the picture to ensure that only exudates. On the other hand, if the parameters do not meet the specified parameters, it will be considered to be contaminated by the noise within the image (Fig. 5).

Fig. 4. Optical disc detection

(a)   **(b)**

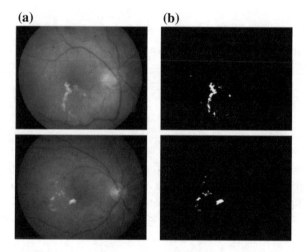

Fig. 5. **a** Examples of original images **b** Examples of exudates extraction

2.4   Exudate Classification

This paper is divided classification into two classes, the classes form the normal retina and abnormal retinal. Classification by the output images of the system, if no exudates in the output image shows normal retina. However, if the images output is assumed to have exudate show retinal disorders.

3   Results and Discussions

System accuracy was measured by all lesions related to DR (exudates). The general precision of the system is measured by its accuracy, which can be determined by the total number of accurate classifications compared to the overall amount of classifications, as detailed in Eq. (1).

$$Accuracy = \frac{TP + TN}{TP + TN + FN + FP} \tag{1}$$

Where TP is detected exudates and fact there are exudates, FN is detected exudates and fact there are non-exudate, TN is undetected exudates in fact non-exudates and FP is undetected exudates but fact there are non-exudates [2]. The evaluation system has been tested by comparing the results of the diagnosis by an ophthalmologist. The system also demonstrates 85% accuracy of exudate detection. In this paper we studied from 60 retina images dividing into the normal retina images 15 images and the abnormal retina image 45 images. The accuracy was measured from percentage of correct classification as the results shown in Table 1.

Table 1. The result of accuracy rate

Class	Number of data test	Number of correct classification	Percentage of correct classification (%)
Normal	15	9	60
Abnormal	45	42	93

Finally, the performances of the proposed method and the Sreng et al. method found in [2] were compared. To compare accuracy of the proposed method with the Sreng et al. method, the correct classification percentage of both methods in Tables 2 show that the proposed method was lower than the Sreng et al. method. The exudates detection of proposed method was 85% from database of 60 retina images. Sreng et al. method show that 91% of exudate is extracted correctly from database of 100 retina images.

Table 2. Comparison of results for exudates detection

Author	Number of retina images	Percentage of correct classification (%)
Sreng et al. [2]	100	91
Our method	60	85

To compare exudates detection steps of the proposed method with the Sreng et al. method, the advantages of proposed method does not even require the circular in optic disc detection process that is able to apply this method on others database of diabetic retinopathy. However, the Sreng et al. method need the circular in optic disc detection process.

4 Conclusions

This paper proposed the detection of diabetic retinopathy using image processing. The process consists of four main parts, including image acquisition, pre-processing, exudate extraction, and exudate classification, respectively. The data sets are separated into two groups for normal retina image and abnormal retina image. The group classified by exudates detection, if detected exudate in the retina image is treated as a class abnormal. But, exudates undetected is treated in class normal. The accuracy results of the system is 85%, indicating that it can help ophthalmologists in finding the thick greasy substance to detect diabetic retinopathy. The system can also be developed for use in medical methods. Nevertheless, this presented method has limitation in optical disk extraction that the optical disk image must be clearly seen. In the future work, we are planning to apply a neural network for exudate classification.

Acknowledgements. This project is supported by Department of Computer Engineering, Faculty of Engineering, Mahidol University. We would like to thank Institute of Medical Research and Technology Assessment for the database.

References

1. Ahmad Zikri, R., Hadzli, H., Syed, F.: A proposed diabetic retinopathy classification algorithm with statistical inference of exudates detection. In: 2013 International Conference, Electrical, Electronics and System Engineering (ICEESE), pp. 80–95. IEEE Press, Malaysia (2013)
2. Sreng, S., Maneerat, N., Isarakorn, D., Pasaya, B., Takada, J., Panjaphongse, R., Varakulsiripunth, R.: Automatic exudate extraction for early detection of diabetic retinopathy. In: International Conference on Information Technology and Electrical Engineering (ICITEE), pp. 31–35. IEEE Press, Thailand (2013)
3. Shahin, E.M., Taha, T.E., Al-Nuaimy, W., El Rabaie, S., Zahran, O.F., El-Samie, F.E.A.: Automated detection of diabetic retinopathy in blurred digital fundus images. In: 8th International Computer Engineering Conference (ICENCO), pp. 20–25. UK (2013)
4. Verma, K., Deep, P., Ramakrishnan, A.G.: Detection and classification of diabetic retinopathy using retinal images, In: India Conference (INDICON), pp. 1–6. IEEE Press, India (2011)
5. Ning, D., Yafen, L.: Automated identification of diabetic retinopathy stages using support vector machine. In: 32nd Chinese on Control Conference (CCC), pp. 3882–3886. IEEE Press, China (2012)
6. Saravanan, V., Venkatalakshmi, B., Farhana, S.M.N.: Design and development of pervasive classifier for diabetic retinopathy. In: 2013 IEEE Conference on Information & Communication Technologies (ICT), pp. 231–235. IEEE Press, India (2013)
7. Zeljkovic, V., Bojic, M., Tameze, C., Valev, V.: Classification algorithm of retina images of diabetic patients based on exudates detection. In: 2012 International Conference on High Performance Computing and Simulation (HPCS), pp. 167–173. IEEE Press, New York (2012)
8. Dhiravidachelvi, E., Rajamani, V.: Computerized detection of optic disc in diabetic retinal images using background subtraction model. In: 2014 International Conference on Circuit, Power and Computing Technologies (ICCPCT), pp. 1217–1222. IEEE Press, India (2014)

Printed in the United States
By Bookmasters